Algebra 2

**LARSON
BOSWELL
KANOLD
STIFF**

Applications • Equations • Graphs

Chapter 2
Resource Book

The Resource Book contains the wide variety of blackline masters available for Chapter 2. The blacklines are organized by lesson. Included are support materials for the teacher as well as practice, activities, applications, and assessment resources.

McDougal Littell
A HOUGHTON MIFFLIN COMPANY
Evanston, Illinois • Boston • Dallas

Contributing Authors

The authors wish to thank the following individuals for their contributions to the Chapter 2 Resource Book.

Rose Elaine Carbone
José Castro
Lee Creighton
John Graham
Fr. Chris M. Hamlett
Edward H. Kuhar
Cheryl A. Leech
Ann C. Nagosky
Karen Ostaffe
Leslie Palmer
Ann Larson Quinn, Ph. D.
Chris Thibaudeau

ISBN: 0-618-02010-1

Contents

Contents

Contents

Descriptions of Resources

This Chapter Resource Book is organized by lessons within the chapter in order to make your planning easier. The following materials are provided:

Tips for New Teachers These teaching notes provide both new and experienced teachers with useful teaching tips for each lesson, including tips about common errors and inclusion.

Parent Guide for Student Success This guide helps parents contribute to student success by providing an overview of the chapter along with questions and activities for parents and students to work on together.

Prerequisite Skills Review Worked-out examples are provided to review the prerequisite skills highlighted on the Study Guide page at the beginning of the chapter. Additional practice is included with each worked-out example.

Strategies for Reading Mathematics The first page teaches reading strategies to be applied to the current chapter and to later chapters. The second page is a visual glossary of key vocabulary.

Lesson Plans and Lesson Plans for Block Scheduling This planning template helps teachers select the materials they will use to teach each lesson from among the variety of materials available for the lesson. The block-scheduling version provides additional information about pacing.

Warm-Up Exercises and Daily Homework Quiz The warm-ups cover prerequisite skills that help prepare students for a given lesson. The quiz assesses students on the content of the previous lesson. (Transparencies also available)

Activity Support Masters These blackline masters make it easier for students to record their work on selected activities in the Student Edition.

Alternative Lesson Openers An engaging alternative for starting each lesson is provided from among these four types: *Application, Activity, Graphing Calculator,* or *Visual Approach.* (Color transparencies also available)

Graphing Calculator Activities with Keystrokes Keystrokes for four models of calculators are provided for each Technology Activity in the Student Edition, along with alternative Graphing Calculator Activities to begin selected lessons.

Practice A, B, and C These exercises offer additional practice for the material in each lesson, including application problems. There are three levels of practice for each lesson: A (basic), B (average), and C (advanced).

Contents

Reteaching with Practice These two pages provide additional instruction, worked-out examples, and practice exercises covering the key concepts and vocabulary in each lesson.

Quick Catch-Up for Absent Students This handy form makes it easy for teachers to let students who have been absent know what to do for homework and which activities or examples were covered in class.

Cooperative Learning Activities These enrichment activities apply the math taught in the lesson in an interesting way that lends itself to group work.

Interdisciplinary Applications/Real-Life Applications Students apply the mathematics covered in each lesson to solve an interesting interdisciplinary or real-life problem.

Math and History Applications This worksheet expands upon the Math and History feature in the Student Edition.

Challenge: Skills and Applications Teachers can use these exercises to enrich or extend each lesson.

Quizzes The quizzes can be used to assess student progress on two or three lessons.

Chapter Review Games and Activities This worksheet offers fun practice at the end of the chapter and provides an alternative way to review the chapter content in preparation for the Chapter Test.

Chapter Tests A, B, and C These are tests that cover the most important skills taught in the chapter. There are three levels of test: A (basic), B (average), and C (advanced).

SAT/ACT Chapter Test This test also covers the most important skills taught in the chapter, but questions are in multiple-choice and quantitative-comparison format. (See *Alternative Assessment* for multi-step problems.)

Alternative Assessment with Rubrics and Math Journal A journal exercise has students write about the mathematics in the chapter. A multi-step problem has students apply a variety of skills from the chapter and explain their reasoning. Solutions and a 4-point rubric are included.

Project with Rubric The project allows students to delve more deeply into a problem that applies the mathematics of the chapter. Teacher's notes and a 4-point rubric are included.

Cumulative Review These practice pages help students maintain skills from the current chapter and preceding chapters.

CHAPTER 2

Tips for New Teachers

For use with Chapter 2

LESSON 2.1

TEACHING TIP Students should be able to find the domain and range of a function from either its mapping, graph, or equation. You might need to teach students how to do so for each of these representations of a function. Make sure to have some examples of functions that have a restricted domain and/or range, such as a quadratic or a square-root function.

TEACHING TIP Show your students the difference between the *mathematical* domain/range of a function and the *real-life* domain/range of that same function in a given word problem. For instance, in Example 6 on page 70, the *mathematical* domain and range of the function $d = 97.8 \cdot t$ are all real numbers. However, its *real-life* domain and range are restricted as shown in the problem. Students need to use common sense to determine the domain and range in word problems.

TEACHING TIP Some students learn how to use the *vertical line test* without understanding how it derives from the definition of *function*. Remind your students that if two points on a graph lie on a vertical line they must have the same x-coordinate, or input, and different y-coordinates, or outputs. Therefore, if a vertical line intersects a graph twice, this means that for a given input there are two different outputs and the relation is not a function.

LESSON 2.2

COMMON ERROR When discussing page 77, students might think that lines are either parallel or perpendicular. Remind them that lines could also be oblique or overlapping by showing them examples and drawings.

LESSON 2.3

COMMON ERROR Some students "plot" the slope, as if it were a point. This way, a slope such as $\frac{4}{3}$ is plotted as the point (3, 4). Remind students that the slope represents *change* in coordinates. This means that when students use the slope to graph a line they must start from a point *on* the line, not the origin.

COMMON ERROR Students might incorrectly interpret the negative sign of a slope to mean "move down *and* left." If they do so, the line will rise from left to right instead of falling. To avoid this mistake, ask your students to decide whether the line will rise or fall *before* they graph it. Then, after they graph it, they should check whether their graph agrees with what they thought.

LESSON 2.4

TEACHING TIP Teach your students how to find the equation of a line given a point on the line and its slope by plugging these values into the equation $y = mx + b$. They will get an equation that they can solve to find the value of the y-intercept, b.

INCLUSION The typical problems asking for the equation of a line parallel or perpendicular to a given line can be too abstract for some students. A visual representation might help them to organize and understand the information from the problem. Students should sketch a graph of the problem showing what they know and what they must find—they can use different colors to differentiate the data from the unknown.

LESSON 2.5

COMMON ERROR Many students believe they must take two data points to draw the best-fitting line and, therefore, to find its equation. Some of these students always take the left-most and the right-most data points on the scatter plot. Remind your students that the best-fitting line might not go through any of the data points. Rather than choosing two points to graph the line, students must graph the line first and then take two points on it to find its equation.

TEACHING TIP Best-fitting lines allow us to make predictions, but their validity is often limited to a small domain. Use Example 3 on page 102 to show students that they must be careful to consider whether the model is valid for the case considered. Ask them to estimate the number of hours slept per day by a 5 year old, and they will get $h \approx 4.9$. Discuss with your students what would be a valid domain for the best-fitting line for this problem as well as other examples you might use.

LESSON 2.6

TEACHING TIP Some students learn how to graph a linear inequality in two variables but are not able to list possible solutions of the inequality. Make sure that your students understand why one of the half-planes is shaded and how to use that graph to find some solutions of the inequality. Ask students whether all points on the shaded region are possible solutions of the inequality and create an example where the solutions must be limited to lattice points (you can link this kind of problem to the ideas of domain and range).

LESSON 2.7

TEACHING TIP If your students do not understand the need to show open or solid dots on a graph at the points of discontinuity, try the following approach. Sketch the graph of a discontinuous piecewise function without writing its corresponding rule. Then ask your students what is the value of the function at the point of discontinuity. Looking at this graph, students will know there are two possible values for the function, but they will not know which one to take. Remind students that a function cannot yield two different outputs for the same input so the graph must clearly show which one is the value of the function. Using a solid and an open dot help us to do so.

COMMON ERROR Some students always connect the different segments and/or rays of a piecewise function, without allowing any points of discontinuity. These graphs are not even functions, because they include vertical segments. Remind students that the different parts of a piecewise function do not have to be connected.

LESSON 2.8

COMMON ERROR Some students always use a table of values to graph functions. This is not wrong but can lead the students to get incorrect answers. For example, if students make a table of values for $-3 \le x \le 3$ for the function $y = |x - 5|$, their graph is most likely to be a line. Ask your students to identify what type of graph they will get beforehand just by looking at the equation.

Outside Resources

BOOKS/PERIODICALS

Weist, Lynda R., and Robert J. Quinn. "Exploring Probability Through an Even-Odds Dice Game." *Mathematics Teaching in the Middle School* (March 1999), pp. 358–362.

ACTIVITIES/MANIPULATIVES

Hadley, William S. "Experiments from Psychology and Neurology." *Activities: Mathematics Teacher* (October 1996), pp. 562–569.

SOFTWARE

Dugdale, Sharon, and David Kibbey. *Green Globs*. Writing equations through given points; challenging game format. Pleasantville, NY: Sunburst Communications.

VIDEOS

Algebra for All and Its Impact. Video of panel discussion on content and teaching of algebra. Reston, VA: NCTM, 1998.

NAME _____ DATE _____

Parent Guide for Student Success

For use with Chapter 2

Chapter Overview One way that you can help your student succeed in Chapter 2 is by discussing the lesson goals in the chart below. When a lesson is completed, ask your student to interpret the lesson goals for you and to explain how the mathematics of the lesson relates to one of the key applications listed in the chart.

Lesson Title	Lesson Goals	Key Applications
2.1: Functions and Their Graphs	Identify and represent relations and functions. Graph linear functions and evaluate functions.	• Ballooning • Boston Marathon • Water Pressure
2.2: Slope and Rate of Change	Find the slope of a line, classify parallel and perpendicular lines, and use slope to solve real-life problems.	• Ladder Safety • Deserts • Oceanography
2.3: Quick Graphs of Linear Equations	Use the slope-intercept form or the standard form of a linear equation to graph the equation.	• Buying a Computer • Rainforests • Car Wash
2.4: Writing Equations of Lines	Write linear equations. Write and use direct variation equations.	• Politics • Jewelry • Breaking Waves
2.5: Correlation and Best-Fitting Lines	Use a scatter plot to identify the correlation shown by a set of data. Approximate the best-fitting line for a set of data.	• Sleep Requirements • Old Faithful • City Year
2.6: Linear Inequalities in Two Variables	Graph linear inequalities in two variables and use linear inequalities to solve real-life problems.	• Communication • Nutrition • Movies
2.7: Piecewise Functions	Represent piecewise functions and use piecewise functions to model real-life quantities.	• Urban Parking • Social Security • Snowstorm
2.8: Absolute Value Functions	Represent absolute value functions and use them to model real-life situations.	• Billiards • Music Singles • Sound Levels

Study Strategy

Making a Skills File is the study strategy featured in Chapter 2 (see page 66). Encourage your student to record the key skills covered in each lesson along with an example that illustrates the skill. Have your student look over the goals and example titles to be sure no skill is missed.

NAME _____ DATE _____

Parent Guide for Student Success

For use with Chapter 2

Key Ideas Your student can demonstrate understanding of key concepts by working through the following exercises with you.

Lesson	Exercise
2.1	Is the relation $(0, 1)$, $(0, -1)$, $(2, 3)$ a function? Explain.
2.2	Ty was 50 in. tall when he was 8 years old and 60 in. tall when he was 12. What was the rate of change in Ty's height?
2.3	Marcia is trying to earn $240 to buy a bicycle. She can earn $8 an hour babysitting or $10 an hour mowing lawns. Write a model for the problem. Find the intercepts of the graph of the model.
2.4	Write an equation of the line that passes through $(3, -1)$ and is perpendicular to the line $y = -3x + 5$.
2.5	Is there usually a positive correlation, a negative correlation, or no correlation between the age of a car and its market value?
2.6	Is $(-5, 6)$ part of the graph of $3x + 2y \leq -1$?
2.7	The cost of x T-shirts at a sale is given by the function below. Find the cost of three T-shirts and of six T-shirts. $C(x) = \begin{cases} 10x, & \text{if } 0 \leq x \leq 2 \\ 8x, & \text{if } 2 < x \leq 4 \\ 5x, & \text{if } x > 4 \end{cases}$
2.8	For $y = -2|x - 5| + 2$, identify the vertex and tell whether the graph opens up or down.

Home Involvement Activity

Directions: Think of a real-life situation that can be represented by a piecewise function, such as the postal rates on page 119. Write equations to model the function and sketch its graph. Check your work by testing a few different values.

Answers

2.1: No; for the input 0 there are two outputs, 1 and -1. **2.2:** 2.5 inches per year
2.3: $8x + 10y = 240$; x-intercept: 30, y-intercept: 24 **2.4:** $y = \frac{1}{3}x - 2$ **2.5:** negative correlation
2.6: yes **2.7:** three T-shirts: $24, six T-shirts: $30 **2.8:** $(5, 2)$; down

NAME _____ DATE _____

Prerequisite Skills Review
For use before Chapter 2

EXAMPLE 1 *Evaluating an Algebraic Expression*

Evaluate the expression for the given values of x and y.

a. $\dfrac{6-y}{7-x}$; $x = -1$, $y = -4$

b. $\dfrac{x-10}{y-4}$; $x = 12$, $y = 2$

SOLUTION

a. $\dfrac{6-y}{7-x} = \dfrac{6-(-4)}{7-(-1)}$ Substitute -1 for x and -4 for y.

$= \dfrac{6+4}{7+1}$ Perform operation.

$= \dfrac{10}{8}$

$= \dfrac{5}{4}$ Simplify.

b. $\dfrac{x-10}{y-4} = \dfrac{(12)-10}{(2)-4}$ Substitute 12 for x and 2 for y.

$= \dfrac{2}{-2}$ Perform operation.

$= -1$ Simplify.

Exercises for Example 1

Evaluate the expression for the given values of x and y.

1. $\dfrac{4-y}{8-x}$; $x = 4$, $y = -8$

2. $\dfrac{x-9}{y-3}$; $x = -3$, $y = 7$

3. $\dfrac{(-4)-y}{2-x}$; $x = -6$, $y = 8$

4. $\dfrac{x-(-3)}{y-4}$; $x = -9$, $y = 5$

EXAMPLE 2 *Rewriting an Equation with More Than One Variable*

Solve the equation for y.

a. $5x + y = 12$

b. $3x - 4y = -8$

SOLUTION

a. $5x + y = 12$ Write original equation.

$y = -5x + 12$ Subtract $5x$ from each side.

b. $3x - 4y = -8$ Write original equation.

$-4y = -3x - 8$ Subtract $3x$ from each side.

$y = \dfrac{3}{4}x + 2$ Divide each side by -4.

NAME _____ DATE _____

Prerequisite Skills Review

For use before Chapter 2

Exercises for Example 2

Solve the equation for y.

5. $-x + 2y = -8$

6. $6x - 3y = 15$

7. $-2x - 7y = -49$

8. $10x + 5y = -75$

EXAMPLE 3 *Solving an Inequality*

Solve the inequality.

a. $-3x + 10 \geq -8$

b. $5y + 12 < -3y - 4$

SOLUTION

a. $-3x + 10 \geq -8$	Write original inequality.
$-3x \geq -18$	Subtract 10 from each side.
$x \leq 6$	Divide each side by -3 and reverse the inequality.
b. $5y + 12 < -3y - 4$	Write original inequality.
$8y + 12 < -4$	Add $3y$ to each side.
$8y < -16$	Subtract 12 from each side.
$y < -2$	Divide each side by 8.

Exercises for Example 3

Solve the inequality.

9. $x + 4 > -8x - 23$

10. $-2y - 6 \geq -4$

11. $4x - 17 \leq -5$

12. $4y + 11 < 6y - 9$

• NAME _____ DATE _____

Strategies for Reading Mathematics

For use with Chapter 2

Strategy: Reading Vocabulary and Taking Notes in Algebra

You have probably already noted how important it is to understand mathematical terms when studying algebra. You can make your studying easier by looking for highlighted vocabulary terms in **heavy type** like this. Then write definitions and examples in your notebook to help you remember the terms. The notebook page below shows a sample of the notes you might take about the vocabulary in the next paragraph.

A **relation** is a mapping, or pairing, of input values with output values. The set of input values is the **domain**, and the set of output values is the **range**. A relation is a **function** provided there is exactly one output for each input. It is not a function if at least one input has more than one output.

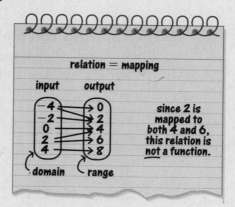

relation = mapping

| input | output |

since 2 is mapped to both 4 and 6, this relation is not a function.

domain   range

STUDY TIP
Reading Vocabulary

When you see highlighted words in heavy type, that means the terms are important. Read the sentence that the term is in for its definition. Check sentences just before and after the term to make sure that you have a complete definition.

STUDY TIP
Taking Notes

Make your notes brief. Write important words that will help you remember the meaning of each term. You might also include pictures, diagrams, or examples.

Questions

1. What are the domain and range of the relation shown on the notebook above? Which other input value could you use to show the relation is not a function?

2. Use a table to represent the relation above. Circle or highlight the parts of your table that show the relation is not a function.

3. What are the domain and range of the relation shown at the right? Is the relation a function? Explain how you know.

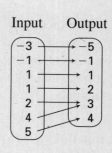

Input   Output

4. The domain of a relation is $-3, -2, -1, 0, 1, 2, 3$, and the relation maps each value to its square. What is the range of the relation? Make a table or diagram to represent the relation. Is this relation a function?

NAME _____ DATE _____

Strategies for Reading Mathematics

For use with Chapter 2

Visual Glossary

The Study Guide on page 66 lists the key vocabulary for Chapter 2 as well as review vocabulary from the previous chapter. Use the page references on page 66 or the Glossary in the textbook to review key terms from the prior chapter. Use the visual glossary below to help you understand some of the key vocabulary in Chapter 2. You may want to copy these diagrams into your notebook and refer to them as you complete the chapter.

GLOSSARY

Linear function (p. 69)

A function of the form $y = mx + b$ where m and b are constants. The graph of a linear function is a line.

Solution of a linear inequality in two variables (p. 108)

An ordered pair (x, y) that, when x and y are substituted in the inequality, gives a true statement.

Graph of a linear inequality in two variables (p. 108) The graph of all solutions of the inequality.

Piecewise function (p. 114)

A function represented by a combination of equations, each corresponding to a part of the domain.

Step function (p. 115) A piecewise function whose graph resembles a set of stair steps.

Graphing the Solution of a Linear Inequality

First, graph the line that forms the boundary of the solution region. Then shade the appropriate half-plane.

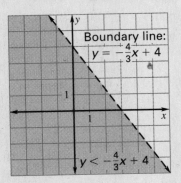

Boundary line:
$y = -\frac{4}{3}x + 4$

$y < -\frac{4}{3}x + 4$

Use a dashed boundary line for $<$.

Check if $(0, 0)$ is a solution.

$$y < -\frac{4}{3}x + 4$$

$$0 \overset{?}{<} -\frac{4}{3} \cdot 0 + 4$$

$$0 < 4 \quad \checkmark$$

Shade the half-plane containing the origin.

Graphing a Piecewise Function

Graph each piece separately. Be careful to graph the correct point wherever there is a jump from one piece to another.

$$f(x) = \begin{cases} 2x, & \text{if } x < 0 \\ -\frac{1}{2}x + 2, & \text{if } x \geq 0 \end{cases}$$

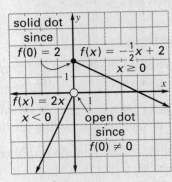

solid dot since $f(0) = 2$

$f(x) = -\frac{1}{2}x + 2$
$x \geq 0$

$f(x) = 2x$
$x < 0$

open dot since $f(0) \neq 0$

TEACHER'S NAME _____ CLASS _____ ROOM _____ DATE _____

Lesson Plan

1-day lesson (See *Pacing the Chapter,* TE pages 64C–64D)

For use with pages 67–74

GOALS 1. **Represent relations and functions.**
2. **Graph and evaluate linear functions.**

State/Local Objectives _____

_____

✓ Check the items you wish to use for this lesson.

STARTING OPTIONS
_____ Prerequisite Skills Review: CRB pages 5–6
_____ Strategies for Reading Mathematics: CRB pages 7–8
_____ Warm-Up or Daily Homework Quiz: TE pages 67 and 56, CRB page 11, or Transparencies

TEACHING OPTIONS
_____ Motivating the Lesson: TE page 68
_____ Lesson Opener (Visual Approach): CRB page 12 or Transparencies
_____ Examples 1–6: SE pages 67–70
_____ Extra Examples: TE pages 68–70 or Transparencies; Internet
_____ Closure Question: TE page 70
_____ Guided Practice Exercises: SE page 71

APPLY/HOMEWORK
Homework Assignment
_____ Basic 20–48 even, 49, 51, 59–62, 65–81 odd
_____ Average 20–48 even, 49, 51–62, 65–81 odd
_____ Advanced 20–48 even, 49–63, 64–81

Reteaching the Lesson
_____ Practice Masters: CRB pages 13–15 (Level A, Level B, Level C)
_____ Reteaching with Practice: CRB pages 16–17 or Practice Workbook with Examples
_____ Personal Student Tutor

Extending the Lesson
_____ Applications (Interdisciplinary): CRB page 19
_____ Challenge: SE page 74; CRB page 20 or Internet

ASSESSMENT OPTIONS
_____ Checkpoint Exercises: TE pages 68–70 or Transparencies
_____ Daily Homework Quiz (2.1): TE page 74, CRB page 23, or Transparencies
_____ Standardized Test Practice: SE page 74; TE page 74; STP Workbook; Transparencies

Notes _____

_____

_____

TEACHER'S NAME _____ CLASS _____ ROOM _____ DATE _____

Lesson Plan for Block Scheduling

Half-day lesson (See *Pacing the Chapter,* TE pages 64C–64D)   For use with pages 67–74

GOALS
1. **Represent relations and functions.**
2. **Graph and evaluate linear functions.**

State/Local Objectives _____

_____

_____

CHAPTER PACING GUIDE	
Day	**Lesson**
1	**2.1 (all)**; 2.2 (all)
2	2.3 (all)
3	2.4 (all)
4	2.5 (all); 2.6 (all)
5	2.7 (all); 2.8 (all)
6	Review/Assess Ch. 2

✓ **Check the items you wish to use for this lesson.**

STARTING OPTIONS
____ Prerequisite Skills Review: CRB pages 5–6
____ Strategies for Reading Mathematics: CRB pages 7–8
____ Warm-Up or Daily Homework Quiz: TE pages 67 and 56,
        CRB page 11, or Transparencies

TEACHING OPTIONS
____ Motivating the Lesson: TE page 68
____ Lesson Opener (Visual Approach): CRB page 12 or Transparencies
____ Examples 1–6: SE pages 67–70
____ Extra Examples: TE pages 68–70 or Transparencies; Internet
____ Closure Question: TE page 70
____ Guided Practice Exercises: SE page 71

APPLY/HOMEWORK
Homework Assignment (See also the assignment for Lesson 2.2.)
____ Block Schedule: 20–48 even, 49, 51–62, 65–81 odd

Reteaching the Lesson
____ Practice Masters: CRB pages 13–15 (Level A, Level B, Level C)
____ Reteaching with Practice: CRB pages 16–17 or Practice Workbook with Examples
____ Personal Student Tutor

Extending the Lesson
____ Applications (Interdisciplinary): CRB page 19
____ Challenge: SE page 74; CRB page 20 or Internet

ASSESSMENT OPTIONS
____ Checkpoint Exercises: TE pages 68–70 or Transparencies
____ Daily Homework Quiz (2.1): TE page 74, CRB page 23, or Transparencies
____ Standardized Test Practice: SE page 74; TE page 74; STP Workbook; Transparencies

Notes _____

_____

_____

NAME _____ DATE _____

WARM-UP EXERCISES

For use before Lesson 2.1, pages 67–74

Evaluate the expression when $x = -2$.

1. $4x - 2$

2. $\dfrac{1}{3}x + \dfrac{5}{3}$

3. $2x^2 - x + 4$

4. $-4x^2 - 6x + 12$

5. $\dfrac{1}{2}|x - 3| - \dfrac{3}{2}$

DAILY HOMEWORK QUIZ

For use after Lesson 1.7, pages 49–56

1. Rewrite $|2 - 5x| = 6$ as two linear equations.

2. Is -15 a solution of $\left|20 - \dfrac{5}{3}x\right| = 5$?

3. Solve $|4 - 6x| = 2$.

4. Rewrite $|5 + 4x| < 7$ as a compound inequality.

Solve the inequality. Then graph your solution.

5. $|9 - 4x| < 5$

6. $|2x + 10| \geq 4$

NAME _____ DATE _____

Visual Approach Lesson Opener

For use with pages 67–74

These are functions.

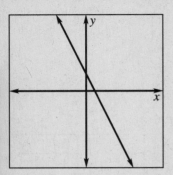 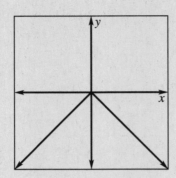 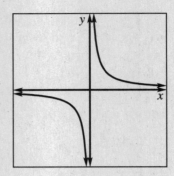

These are not functions.

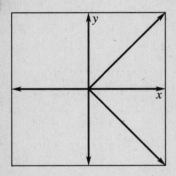 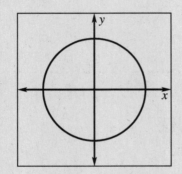 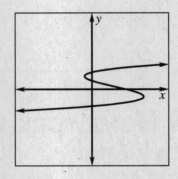

Four of the following are functions. Which are they?

A. 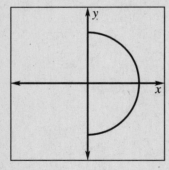 **B.** 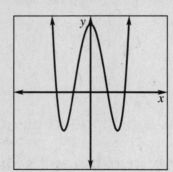 **C.**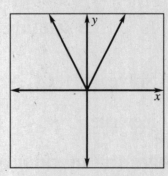

D. 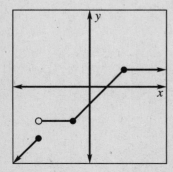 **E.** 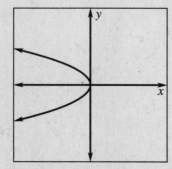 **F.**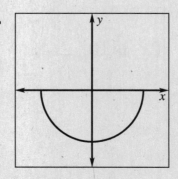

NAME _____ DATE _____

Practice A

For use with pages 67–74

Identify the domain and range.

1. Input   Output

2 ——→ 3
−1 ——→ 6
0 ——→ 16

2. Input   Output

3 ——→ −9
9 ——→ 0
4

3. Input   Output

1 ——→ −12
2 ——→ 24
——→ 6

Graph the relation. Then tell whether the relation is a function.

4.

x	0	1	2	3	4
y	3	5	3	1	0

5.

x	−3	−3	4	5	9
y	0	1	−2	3	11

6.

x	−2	−1	0	1	2
y	1	2	3	4	5

Use the vertical line test to determine whether the relation is a function.

7.
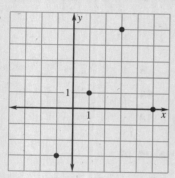

8.
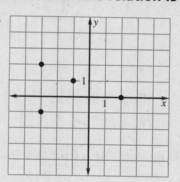

9.
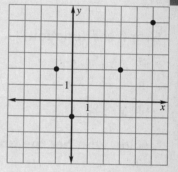

Complete the table of values for the given function. Then graph the function.

10. $y = 2x + 3$

x	−2	−1	0	1	2
y					

11. $y = -\frac{1}{2}x + 4$

x	−2	−1	0	1	2
y					

Graph the function.

12. $y = x - 2$

13. $y = -x + 3$

14. $y = 3x + 4$

15. $y = -6x + 2$

16. $y = 4x - 3$

17. $y = -3x - 5$

18. $y = 8x$

19. $y = -2$

20. $y = \frac{1}{2}x + 5$

21. **U.S. Open Champions**   The table shows the golf scores of the U.S. Open Champions from 1986 to 1996. Use a coordinate plane to graph these results.

Year	1986	1987	1988	1989	1990	1991	1992	1993	1994	1995	1996
Score	279	285	281	283	278	277	275	277	279	274	276

NAME _____ DATE _____

Practice B

For use with pages 67–74

Graph the relation. Then tell whether the relation is a function.

1.

x	−2	−1	0	1	2
y	0	5	6	0	3

2.

x	−2	−1	0	1	2	−2
y	4	−1	3	2	1	−8

Use the vertical line test to determine whether the relation is a function.

3.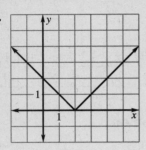

4.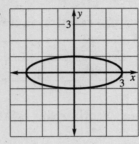

5.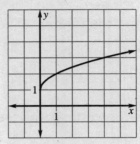

Graph the function.

6. $y = 5x + 1$

7. $y = 3x - 7$

8. $y = -2x$

9. $y = -x + 2$

10. $y = \frac{1}{2}x + 3$

11. $y = -3x - 5$

12. $y = 2x + 3$

13. $y = 2$

14. $y = -\frac{1}{3}x + 1$

Decide whether the function is linear. Then find the indicated value of $f(x)$.

15. $f(x) = x + 7; \ f(-3)$

16. $f(x) = x^3 - x + 2; \ f(1)$

17. $f(x) = 4 - 3x; \ f(2)$

18. $f(x) = |3x + 1|; \ f(-5)$

19. $f(x) = \dfrac{3}{x + 2}; \ f(4)$

20. $f(x) = \dfrac{3}{4}x - 1; \ f(8)$

21. **Geometry** The surface area of a cube with side length x is given by the function $S(x) = 6x^2$. Find $S(3)$. Explain what $S(3)$ represents.

Statistics   In Exercises 22–24, use the following information.
The table below shows the number of games won and lost by the teams in the
Eastern Division of the NFL's National Football Conference for the 1996 season.

Team	Won, x	Lost, y
Dallas Cowboys	10	6
Philadelphia Eagles	10	6
Washington Redskins	9	7
Arizona Cardinals	7	9
New York Giants	6	10

22. What is the domain of the relation?

23. What is the range of the relation?

24. Is the number of wins a function of the number of losses?

NAME _____ DATE _____

Practice C

For use with pages 67–74

Tell whether the relation is a function.

1.
Input   Output

3 ⟶ −1
2 ⟶ −5
5 ⟶ 5
4 ⟶ 6

2.

x	1	2	4	7	0
y	0	0	0	0	0

3.

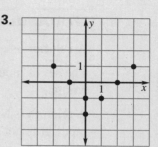

State the quadrant in which each point lies. Assume that a and b are positive numbers.

4. (a, b) **5.** $(-a, b)$ **6.** $(-a, -b)$ **7.** $(a, -b)$

Graph the function.

8. $y = 3x + 5$ **9.** $y = -3$ **10.** $y = 4 - 7x$

11. $y = \frac{1}{2}x + 2$ **12.** $y = 4 - \frac{3}{4}x$ **13.** $y = \frac{3}{5}x$

Decide whether the function is linear. Then find the indicated value of $f(x)$.

14. $f(x) = 7x + 2, \ f(2)$ **15.** $f(x) = x^2 + 3x - 1, \ f(-3)$ **16.** $f(x) = |x| + x, \ f(-5)$

17. $f(x) = (x + 3)^2, \ f(4)$ **18.** $f(x) = \frac{x - 7}{3x}, \ f(2)$ **19.** $f(x) = 2x^3 - 4, \ f(1)$

Earthquakes **In Exercises 20–22, use the table below which shows 10 of the worst earthquakes of the 20th century.**

Location (Year)	Magnitude, x	Deaths, y
Chile (1960)	8.3	5000
India (1950)	8.7	1530
Japan (1946)	8.4	2000
Chile (1939)	8.3	28,000
India (1934)	8.4	10,700
Japan (1933)	8.9	2990
China (1927)	8.3	200,000
Japan (1923)	8.3	200,000
China (1920)	8.6	100,000
Chile (1906)	8.6	20,000

20. Identify the domain and range of the relation.

21. Graph the relation.

22. Is the number of deaths a function of the magnitude of an earthquake? Explain.

NAME _____ DATE _____

Reteaching with Practice

For use with pages 67–74

GOAL Represent relations and functions and graph and evaluate linear functions

---

VOCABULARY

A **relation** is a mapping, or pairing, of input values with output values.

The set of input values is the **domain,** and the set of output values is the **range.**

A relation is a **function** provided there is exactly one output for each input. It is not a function if at least one input has more than one output.

The first number in an ordered pair is the **x-coordinate,** and the second number is the **y-coordinate.**

A **coordinate plane** is formed by two real number lines that intersect at a right angle.

A **quadrant** is one of the four parts into which the axes divide a coordinate plane.

The horizontal axis in a coordinate plane is the **x-axis,** and the vertical axis is the **y-axis.**

The **origin** $(0, 0)$ is the point in a coordinate plane where the axes intersect.

An ordered pair (x, y) is a **solution** of an equation if the equation is true when the values of x and y are substituted into the equation.

The input variable is called the **independent variable,** and the output variable is called the **dependent variable.**

EXAMPLE 1 *Identifying Functions*

Identify the domain and range. Then tell whether the relation is a function.

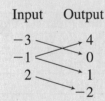

SOLUTION

The domain consists of -3, -1, and 2, and the range consists of $-2, 0, 1,$ and 4. The relation is not a function because the input -1 is mapped onto both 1 and 4.

Exercises for Example 1

Identify the domain and range. Then tell whether the relation is a function.

1. Input   Output
   $-1 \longrightarrow 5$
   $-3 \longrightarrow 0$
   $2 \longrightarrow 2$
   1

2. Input   Output
   $-2 \longrightarrow 6$
   $-3 \longrightarrow 7$
   $5 \longrightarrow 8$

3. Input   Output

   $-4 \longrightarrow 3$
   $-2 \longrightarrow -1$
   $0 \longrightarrow -2$
   4

Reteaching with Practice

For use with pages 67–74

EXAMPLE 2 *Graphing a Function*

Graph the function $y = -\frac{1}{2}x + 1$.

SOLUTION

1. Construct a table of values.

x	-4	-2	0	2	4
y	3	2	1	0	-1

2. Plot the points.

3. Draw a line through the points.

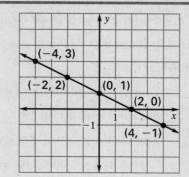

Exercises for Example 2

Graph the function.

4. $y = x - 4$ **5.** $y = -x$ **6.** $y = -3$

7. $y = 3x - 2$ **8.** $y = -x + 4$ **9.** $y = 2x + 2$

EXAMPLE 3 *Evaluating Functions*

Evaluate the function when $x = -4$. The symbol $f(x)$ is an example of **function notation** and is read as "the value of f at x" or "f of x."

a. $f(x) = 5 - 2x$ **b.** $f(x) = x^3 - 9$

SOLUTION

a. $f(x) = 5 - 2x$ Write function.

     $f(-4) = 5 - 2(-4)$ Substitute -4 for x.

          $= 13$ Simplify.

b. $f(x) = x^3 - 9$

     $f(-4) = (-4)^3 - 9$

          $= -73$

Exercises for Example 3

Evaluate the function for the given value of x.

10. $f(x) = 3x - 11; f(0)$ **11.** $f(x) = x^2 - 4; f(1)$ **12.** $f(x) = 6 - x; f(-5)$

NAME _____ DATE _____

Quick Catch-Up for Absent Students

For use with pages 67–74

The items checked below were covered in class on (date missed) _____

Lesson 2.1: Functions and Their Graphs

_____ **Goal 1:** Represent relations and functions. (pp. 67–68)

Material Covered:

_____ Example 1: Identifying Functions

_____ Student Help: Study Tip

_____ Student Help: Skills Review

_____ Example 2: Graphing Relations

_____ Example 3: Using the Vertical Line Test in Real Life

Vocabulary:

relation, p. 67   domain, p. 67

range, p. 67   function, p. 67

ordered pair, p. 67   coordinate plane, p. 67

_____ **Goal 2:** Graph and evaluate linear functions. (pp. 69–70)

Material Covered:

_____ Example 4: Graphing a Function

_____ Student Help: Study Tip

_____ Example 5: Evaluating Functions

_____ Example 6: Using a Function in Real Life

Vocabulary:

equation in two variables, p. 69   solution of an equation in two variables, p. 69

independent variable, p. 69   dependent variable, p. 69

graph of an equation in two variables, p. 69

linear functions, p. 69   function notation, p. 69

_____ Other (specify) _____

_____

Homework and Additional Learning Support

_____ Textbook (specify) <u>pp. 71–74</u> _____

_____

_____ Internet: Extra Examples at www.mcdougallittell.com

_____ *Reteaching with Practice* worksheet (specify exercises) _____

_____ *Personal Student Tutor* for Lesson 2.1

NAME _____ DATE _____

Interdisciplinary Application

For use with pages 67–74

Nutrition

HEALTH   Nutritionists examine many common foods to determine the content of several important substances. For people on restricted diets, this information is vital to healthy living, and food producers are required by law to label their foods with these numbers.

For example, many people monitor their intake of fats and sodium. Sources of these ingredients in everyday meals include cooking oils, salad dressings, and other condiments. The amounts of calories, fat, saturated fat, and sodium for one tablespoon of several of these items is given in the following table:

Type	*Calories*	*Fat (g)*	*Saturated Fat (g)*	*Sodium (mg)*
Butter (salted)	100	11	7.1	116
Margarine (salted)	100	11	2.2	132
Olive Oil	125	14	1.9	0
Bleu Cheese Dressing	75	8	1.5	164
French Dressing (Regular)	85	9	1.4	188
French Dressing (Low Calorie)	25	2	0.2	306
Italian Dressing	80	9	1.3	162
Mayonnnaise	100	11	1.7	80

1. First, consider the columns labeled *Fat* and *Saturated Fat*.

   a. Graph the data in the two columns with *Fat* on the *x*-axis and *Saturated Fat* on the *y*-axis.

   b. Is this graph the graph of a function? Explain why or why not.

   c. Do you notice any points that seem to "stick out" from the rest of the data? Which types of food do these points represent?

   d. Suppose you were cooking a chicken dish for someone who wants their saturated fat intake to be less than two grams. Which of these items would be okay to include in the recipe?

2. Consider the relationship between *Sodium* and *Calories*.

   a. Graph the data with *Calories* on the *x*-axis and *Sodium* on the *y*-axis.

   b. Is this graph the graph of a function? Explain why or why not.

   c. Do you notice any points that seem to "stick out" from the rest of the data? Which types of food do these points represent?

   d. Suppose you were supervising the diet of clients who needed to reduce their sodium intake. If you were to recommend two items from the list, which two would they be?

NAME _____ DATE _____

Challenge: Skills and Applications

For use with pages 67–74

In Exercises 1–5, use the following information.

A function f may have the property that, for *all* numbers a and b in the domain of f,

$$f(a) + f(b) = f(a + b).$$

Check whether each of the following functions has the property described above. If it does not, give an example to show that it does not.

1. $f(x) = x^2$
                                  2. $f(x) = 2x$

3. $f(x) = |x|$
                                  4. $f(x) = 5x + 1$

5. What must be true about the numbers m and/or b in order for a function of the form $f(x) = mx + b$ to have the property above?

In Exercises 6–10, use the following information.

A function with the property that $f(-x) = f(x)$, for all numbers x in its domain, is called an *even* function. A function with the property that $f(-x) = -f(x)$, for all x, is called an *odd* function. Tell whether each of the following is odd or even or neither.

6. $f(x) = x^2 + 2$
                             7. $x + 1$

8. $f(x) = f(x) = x^3 + x$
              9. $f(x) = |x|$

10. If $f(x)$ is a function of the form $f(x) = ax^n + bx^{n-1} + \ldots + nx + p$, what do you think must be true about the coefficients, $a, b, \ldots, p$ in order for f to be an even function? an odd function?

In Exercises 11–14, use the following information.

A *one-to-one* function f with the property that is a and b are any two distinct numbers in the domain of f, $f(a) \neq f(b)$. (In other words, f never maps two distinct inputs to the same output.) Tell whether each of the following functions is one-to-one. If it is not, give an example of a violation of the foregoing definition.

11. $f(x) = x^3$
                                12. $f(x) = x^2 + 1$

13. $f(x) = |x|$
                                14. $f(x) = -2x + 5$

TEACHER'S NAME _____ CLASS _____ ROOM _____ DATE _____

Lesson Plan

1-day lesson (See *Pacing the Chapter,* TE pages 64C–64D)   **For use with pages 75–81**

GOALS   1. **Find slopes of lines and classify parallel and perpendicular lines.**
2. **Use slope to solve real-life problems.**

State/Local Objectives _____

_____

✓ Check the items you wish to use for this lesson.

STARTING OPTIONS
_____ Homework Check: TE page 71; Answer Transparencies
_____ Warm-Up or Daily Homework Quiz: TE pages 75 and 74, CRB page 23, or Transparencies

TEACHING OPTIONS
_____ Motivating the Lesson: TE page 76
_____ Lesson Opener (Activity): CRB page 24 or Transparencies
_____ Graphing Calculator Activity with Keystrokes: CRB pages 25–26
_____ Examples 1–6: SE pages 75–78
_____ Extra Examples: TE pages 76–78 or Transparencies; Internet
_____ Closure Question: TE page 78
_____ Guided Practice Exercises: SE page 79

APPLY/HOMEWORK
Homework Assignment
_____ Basic 17–36, 38–48 even, 49, 51, 54, 59–71 odd
_____ Average 17–47, 48–56 even, 59–71 odd
_____ Advanced 17–50, 52–58, 59–71 odd

Reteaching the Lesson
_____ Practice Masters: CRB pages 27–29 (Level A, Level B, Level C)
_____ Reteaching with Practice: CRB pages 30–31 or Practice Workbook with Examples
_____ Personal Student Tutor

Extending the Lesson
_____ Cooperative Learning Activity: CRB page 33
_____ Applications (Real-Life): CRB page 34
_____ Challenge: SE page 81; CRB page 35 or Internet

ASSESSMENT OPTIONS
_____ Checkpoint Exercises: TE pages 76–78 or Transparencies
_____ Daily Homework Quiz (2.2): TE page 81, CRB page 38, or Transparencies
_____ Standardized Test Practice: SE page 81; TE page 81; STP Workbook; Transparencies

Notes _____

_____

_____

TEACHER'S NAME _____ CLASS _____ ROOM _____ DATE _____

Lesson Plan for Block Scheduling

Half-day lesson (See *Pacing the Chapter,* TE pages 64C–64D)   **For use with pages 75–81**

GOALS
1. **Find slopes of lines and classify parallel and perpendicular lines.**
2. **Use slope to solve real-life problems.**

State/Local Objectives _____

_____

_____

✓ **Check the items you wish to use for this lesson.**

STARTING OPTIONS
____ Homework Check: TE page 71; Answer Transparencies
____ Warm-Up or Daily Homework Quiz: TE pages 75 and 74,
       CRB page 23, or Transparencies

TEACHING OPTIONS
____ Motivating the Lesson: TE page 76
____ Lesson Opener (Activity): CRB page 24 or Transparencies
____ Graphing Calculator Activity with Keystrokes: CRB pages 25–26
____ Examples 1–6: SE pages 75–78
____ Extra Examples: TE pages 76–78 or Transparencies; Internet
____ Closure Question: TE page 78
____ Guided Practice Exercises: SE page 79

APPLY/HOMEWORK
Homework Assignment (See also the assignment for Lesson 2.1.)
____ Block Schedule: 17–47, 48–56 even, 59–71 odd

Reteaching the Lesson
____ Practice Masters: CRB pages 27–29 (Level A, Level B, Level C)
____ Reteaching with Practice: CRB pages 30–31 or Practice Workbook with Examples
____ Personal Student Tutor

Extending the Lesson
____ Cooperative Learning Activity: CRB page 33
____ Applications (Real Life): CRB page 34
____ Challenge: SE page 81; CRB page 35 or Internet

ASSESSMENT OPTIONS
____ Checkpoint Exercises: TE pages 76–78 or Transparencies
____ Daily Homework Quiz (2.2): TE page 81, CRB page 38, or Transparencies
____ Standardized Test Practice: SE page 81; TE page 81; STP Workbook; Transparencies

Notes _____

_____

_____

CHAPTER PACING GUIDE	
Day	Lesson
1	2.1 (all); **2.2 (all)**
2	2.3 (all)
3	2.4 (all)
4	2.5 (all); 2.6 (all)
5	2.7 (all); 2.8 (all)
6	Review/Assess Ch. 2

WARM-UP EXERCISES

For use before Lesson 2.2, pages 75–81

Evaluate the expression.

1. $\dfrac{-1 - 4}{4 - (-3)}$

2. $\dfrac{0 - (-2)}{-1 - (-2)}$

3. $\dfrac{3 - 4}{2 + 1}$

4. $\dfrac{2 - (-2)}{3 - 6}$

DAILY HOMEWORK QUIZ

For use after Lesson 2.1, pages 67–74

1. Identify the domain and range.

Input   Output

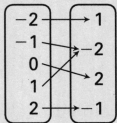

2. Graph the relation in Exercise 1. Then use the vertical line test to tell whether the relation is a function.

3. Graph the function $y = 2x - 1$.

4. Decide whether the function $f(x) = 4x^2 - x + 1$ is linear. Find the value of $f(3)$.

Activity Lesson Opener

For use with pages 75–81

SET UP: Work in a group.
YOU WILL NEED: • colored pencils • ruler

The slope of a line that is not vertical is the ratio of the rise (vertical change) to the run (horizontal change).

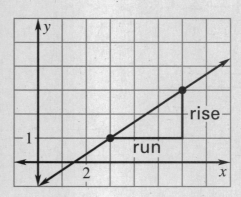

1. What is the slope of the line in the diagram?

2. Plot the points $(2, -2)$ and $(4, -1)$ in the grid below and draw a line through the points.

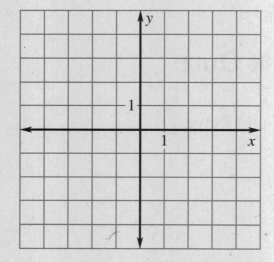

3. Use a colored pencil to mark the rise. Use a different colored pencil to mark the run. Find the slope of the line.

4. Plot the points $(-2, 1)$ and $(2, 3)$ on the same grid. Draw a line through the points. Use colored pencils to find the slope. Write the slope in simplest form.

5. What do you notice about the lines? What do you notice about their slopes?

NAME _____ DATE _____

Graphing Calculator Activity

For use with pages 75-81

GOAL **To classify lines using slopes**

Recall that $y = mx + b$ is a linear function, where m and b are constants. The **slope** of a nonvertical line is the ratio of vertical change (the *rise*) to horizontal change (the *run*). The slope of a line is represented by the letter m.

Activity

① Use a graphing calculator to graph the following.

$y = 0.5x$

$y = x$

$y = 3x$

$y = -3x$

$y = -x$

$y = -0.5x$

② Would you describe the lines with slopes of $m = 0.5$, $m = 1$, and $m = 3$ as *rising from left to right* or falling *from left to right*?

③ Would you describe the lines with slopes of $m = -0.5$, $m = -1$, and $m = -3$ as *rising from left to right* or *falling from left to right*?

④ Choose two lines from Step 1. Compare the absolute values of the slopes. Which line is steeper?

Exercises

1. Determine which of the following are lines that rise from left to right.

a. $y = 2x$ **b.** $y = -5x$ **c.** $y = -0.3$ **d.** $y = 7x$

2. Determine which lines in Exercise 1 fall from left to right.

3. Use a graphing calculator to check your answers to Exercises 1 and 2.

4. Order the following lines in terms of steepness, listing the steepest line first.

$y = 0.3$

$y = 3x$

$y = 7x$

$y = 0.7x$

$y = 5x$

5. Use a graphing calculator to check your answer to Exercise 4.

Graphing Calculator Activity

For use with pages 75–81

TI-82

| Y= | 0.5 | X,T,θ | ENTER |

| X,T,θ | ENTER |

3 | X,T,θ | ENTER |

| (-) | 3 | X,T,θ | ENTER |

| (-) | X,T,θ | ENTER |

| (-) | 0.5 | X,T,θ | ENTER |

| ZOOM | 6 |

TI-83

| Y= | 0.5 | X,T,θ,n | ENTER |

| X,T,θ,n | ENTER |

3 | X,T,θ,n | ENTER |

| (-) | 3 | X,T,θ,n | ENTER |

| (-) | X,T,θ,n | ENTER |

| (-) | 0.5 | X,T,θ,n | ENTER |

| ZOOM | 6 |

SHARP EL-9600c

| Y= | 0.5 | X/θ/T/n | ENTER |

| X/θ/T/n | ENTER |

3 | X/θ/T/n | ENTER |

| (-) | 3 | X/θ/T/n | ENTER |

| (-) | X/θ/T/n | ENTER |

| (-) | 0.5 | X/θ/T/n | ENTER |

| ZOOM | [A] 5 |

CASIO CFX-9850GA PLUS

From the main menu, select GRAPH.

| 0.5 | X,θ,T | EXE |

| X,θ,T | EXE |

3 | X,θ,T | EXE |

| (-) | 3 | X,θ,T | EXE |

| (-) | X,θ,T | EXE |

| (-) | 0.5 | X,θ,T | EXE |

| SHIFT | F3 | F3 | EXIT | F6 |

Lesson 2.2

NAME _____ DATE _____

Practice A

For use with pages 75–81

Estimate the slope of the line.

1.

2.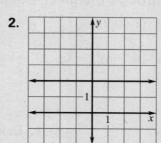

3.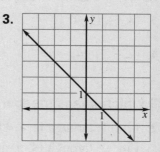

Find the slope of the line passing through the given points.

4. $(2, 3), (5, 9)$

5. $(1, 4), (3, -2)$

6. $(-2, 7), (-3, -1)$

7. $(5, -1), (-7, 5)$

8. $(-11, 0), (4, -5)$

9. $(3, 4), (0, 0)$

Decide whether the line with the given slope *rises, falls, is horizontal,* or *is vertical.*

10. $m = 2$

11. $m = 0$

12. $m = -7$

13. $m = \dfrac{2}{3}$

14. $m = -\dfrac{4}{5}$

15. m is undefined.

Tell whether the lines with the given slopes are *parallel, perpendicular,* or *neither.*

16. Line 1: $m = 2$

Line 2: $m = -2$

17. Line 1: $m = 5$

Line 2: $m = \dfrac{1}{5}$

18. Line 1: $m = -\dfrac{3}{8}$

Line 2: $m = \dfrac{8}{3}$

19. Line 1: $m = 4$

Line 2: $m = 4$

20. Line 1: $m = \dfrac{1}{3}$

Line 2: $m = -3$

21. Line 1: $m = \dfrac{2}{3}$

Line 2: $m = -\dfrac{2}{3}$

22. *Picking Strawberries* One afternoon your family goes out to pick strawberries. At 1:00 P.M., your family has picked 3 quarts. Your family finishes picking at 3:00 P.M. and has 28 quarts of strawberries. At what rate was your family picking?

23. *Ramp* The specifications of a ramp that leads onto a loading dock state that the slope of the ramp must be no steeper than $\frac{1}{64}$. If the ramp begins 200 feet from the base of the loading dock and the dock is 3 feet tall, does the ramp's slope meet the specification?

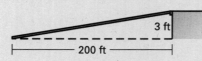

Practice B

For use with pages 75–81

Find the slope of the line passing through the given points.

1. $(4, 5), (2, 9)$ **2.** $(1, 4), (5, 0)$ **3.** $(-3, 5), (6, 2)$

4. $(2, -7), (4, -4)$ **5.** $(0, -8), (-3, -5)$ **6.** $\left(\frac{1}{2}, \frac{3}{4}\right), \left(\frac{3}{2}, \frac{9}{4}\right)$

Tell which line is steeper.

7. Line 1: through $(-2, 1)$ and $(3, 6)$
Line 2: through $(4, 5)$ and $(2, -3)$

8. Line 1: through $(3, -1)$ and $(5, -5)$
Line 2: through $(-2, -2)$ and $(1, -11)$

9. Line 1: through $(0, 3)$ and $(-2, 4)$
Line 2: through $(-8, 6)$ and $(4, -6)$

10. Line 1: through $(10, 2)$ and $(-5, -3)$
Line 2: through $(4, -1)$ and $(12, 0)$

**Find the slope of the line passing through the given points. Then
tell whether the line *rises, falls, is horizontal,* or *is vertical*.**

11. $(4, -2)$ and $(3, -3)$ **12.** $(9, -2)$ and $(-3, -2)$ **13.** $(-3, 5)$ and $(5, 3)$

14. $(7, 5)$ and $(7, -8)$ **15.** $(10, 5)$ and $(4, 15)$ **16.** $(0, 4)$ and $(-3, 4)$

Tell whether the lines are *parallel, perpendicular,* or *neither*.

17. Line 1: through $(3, 2)$ and $(1, 5)$
Line 2: through $(-1, 6)$ and $(2, 8)$

18. Line 1: through $(-3, -1)$ and $(4, -8)$
Line 2: through $(5, 3)$ and $(4, 2)$

19. Line 1: through $(-2, 1)$ and $(-5, 3)$
Line 2: through $(0, 3)$ and $(3, 5)$

20. Line 1: through $(0, 6)$ and $(-5, 0)$
Line 2: through $(-4, 4)$ and $(2, -1)$

21. *Mountainside* The halfway point of a tunnel through a mountain is
$\frac{3}{2}$ miles from either end of the tunnel. The mountain is 660 feet $\left(\frac{1}{8} \text{ mile}\right)$
high. Find the slope of the side of the mountain.

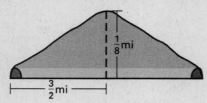

22. *Prom Tickets* You volunteered to take a shift selling prom tickets during
your morning study hall. When your shift began at 11:00 A.M., 50 tickets
had been sold. At 11:40 A.M., when your shift ended, 84 tickets had been
sold. At what rate did you sell prom tickets?

Practice C

For use with pages 75–81

Find the slope of the line passing through the given points.

1. $(6, 3), (-4, -1)$

2. $(-5, -3), (-7, -6)$

3. $\left(\frac{1}{5}, \frac{3}{5}\right), \left(\frac{3}{4}, -\frac{1}{4}\right)$

4. $\left(-1, \frac{1}{3}\right), \left(3, -\frac{2}{3}\right)$

5. $\left(-\frac{3}{5}, -3\right), \left(-\frac{6}{5}, 0\right)$

6. $(-5, 2), (12, -14)$

Decide whether the line passing through the given points *rises, falls, is horizontal*, or *is vertical*.

7. $(-9, -11), (-5, 5)$

8. $(-1, 6), (-1, 7)$

9. $(7, 0), (1, 12)$

Determine which line is steeper.

10. Line 1: through $(3, 7)$ and $(6, 2)$

Line 2: through $(2, 4)$ and $(3, 8)$

11. Line 1: through $(1, 1)$ and $(0, 2)$

Line 2: through $(-1, -4)$ and $(2, -2)$

12. Line 1: through $(5, 2)$ and $(-1, 3)$

Line 2: through $(-3, 4)$ and $(2, 5)$

13. Line 1: through $(-6, 2)$ and $(1, -1)$

Line 2: through $(4, 3)$ and $(-1, 3)$

14. *Parallel Lines*   If two nonvertical lines are parallel, what do you know about their slopes?

15. *Perpendicular Lines*   If two nonvertical lines are perpendicular, what do you know about their slopes?

16. *Vertical Lines*   All vertical lines are parallel to what type of line?

17. *Vertical Lines*   All vertical lines are perpendicular to what type of line?

18. *Washington Monument*   The Washington Monument is 555 feet tall. The monument is composed of a 500-foot pillar topped by a 55-foot pyramid. The base of the pillar is 55 feet wide. The base of the pyramid is 34 feet wide. Approximate the slope of the sides of the pillar and the slope of the pyramid.

19. *Pyramids of Egypt*   The sides of the base of the largest pyramid, Khufu, has length 755 feet. The height of Khufu was originally 482 feet, but now is approximately 450 Feet. Find the slope of a side of the pyramid at its original size and at its present size.

20. *Equilateral Triangles*   An equilateral triangle has the same side lengths and angle measures. Draw an equilateral triangle on a coordinate plane such that one of the vertices is the origin. Approximate the slopes of the sides of your triangle. What are the slopes of the sides of any equilateral triangle in this position?

NAME _____ DATE _____

Reteaching with Practice

For use with pages 75–81

GOAL Find slopes of lines, classify parallel and perpendicular lines, and use slope to solve real-life problems

---

VOCABULARY

The **slope** of a nonvertical line is the ratio of the vertical change (the *rise*) to the horizontal change (the *run*).

Two lines in a plane are **parallel** if they do not intersect and they have the same slope.

Two lines in a plane are **perpendicular** if they intersect to form a right angle. The slopes of two perpendicular lines are negative reciprocals.

EXAMPLE 1 *Finding the Slope of a Line*

Find the slope of the line passing through $(-1, 4)$ and $(-3, 4)$.

SOLUTION

Let $(x_1, y_1) = (-1, 4)$ and $(x_2, y_2) = (-3, 4)$.

$$m = \frac{y_2 - y_1}{x_2 - x_1} \qquad \text{Write formula for slope.}$$

$$= \frac{4 - 4}{-3 - (-1)} \qquad \text{Substitute values.}$$

$$= \frac{0}{-2} \qquad \text{Simplify.}$$

Because the slope is zero, the line is horizontal.

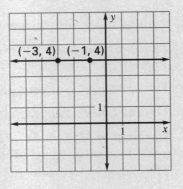

Exercises for Example 1

Find the slope of the line passing through the given points. Then tell whether the line *rises, falls, is horizontal,* or *is vertical.*

1. $(3, 2), (5, 4)$ **2.** $(4, 1), (-6, 3)$ **3.** $(2, 5), (-7, 5)$

4. $(5, -1), (5, 6)$ **5.** $(4, -1), (4, 4)$ **6.** $(2, 5), (-2, -1)$

EXAMPLE 2 *Classifying Parallel and Perpendicular Lines*

Tell whether the lines are *parallel, perpendicular,* or *neither.*

a. Line 1: through $(-1, 3)$ and $(1, -3)$
   Line 2: through $(3, 0)$ and $(0, -1)$

b. Line 1: through $(-2, 6)$ and $(0, -2)$
   Line 2: through $(-1, 6)$ and $(1, -2)$

NAME _____ DATE _____

Reteaching with Practice

For use with pages 75–81

SOLUTION

a. The slopes of the two lines are:

$$m_1 = \frac{3 - (-3)}{-1 - 1} = \frac{6}{-2} = -3$$

$$m_2 = \frac{0 - (-1)}{3 - 0} = \frac{1}{3}$$

Because $m_1 m_2 = -3 \cdot \dfrac{1}{3} = -1$, the lines are

perpendicular.

b. The slopes of the two lines are:

$$m_1 = \frac{6 - (-2)}{-2 - 0} = \frac{8}{-2} = -4$$

$$m_2 = \frac{6 - (-2)}{-1 - 1} = \frac{8}{-2} = -4$$

Because $m_1 = m_2 = -4$, the lines are parallel.

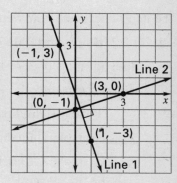

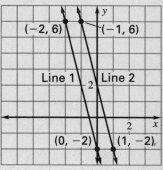

Exercises for Example 2

Tell whether the lines are *parallel*, *perpendicular*, or *neither*.

7. Line 1: through $(2, 4)$ and $(-1, -5)$

Line 2: through $(0, -2)$ and $(-3, 7)$

8. Line 1: through $(-3, 5)$ and $(-6, -1)$

Line 2: through $(-2, -5)$ and $(4, -8)$

EXAMPLE 3 *Slope as a Rate of Change*

Six years ago a house was purchased for $89,000. This year it was
appraised at $125,000. Find the average rate of change and use it to
determine the value after 9 years.

SOLUTION

$$\text{Average rate of change} = \frac{\text{Change in value}}{\text{Change in time}}$$

$$= \frac{\$125,000 - \$89,000}{6 \text{ yrs}} = \frac{\$36,000}{6 \text{ yrs}} = \$6000 \text{ per year}$$

Over a 9-year period, the value changed 9($6000) = $54,000. The value
after 9 years was $89,000 + $54,000 = $143,000.

Exercises for Example 3

9. On a typical summer evening the temperature was 90°F at 7:00 P.M.
At 8:00 A.M. the next morning the temperature was 70.5°F. What is
the average rate of change in temperature? Estimate what the
temperature was at midnight.

NAME _____ DATE _____

Quick Catch-Up for Absent Students

For use with pages 75–81

The items checked below were covered in class on (date missed) _____ .

Lesson 2.2: Slope and Rate of Change

____ **Goal 1**: Find slopes of lines and classify parallel and perpendicular lines. (pp. 75–77)

Material Covered:

____ Student Help: Look Back

____ Example 1: Finding the Slope of a Line

____ Example 2: Classifying Lines Using Slope

____ Student Help: Study Tip

____ Example 3: Comparing Steepness of Lines

____ Example 4: Classifying Parallel and Perpendicular Lines

Vocabulary:

slope, p. 75   parallel lines, p. 77
perpendicular lines, p. 77

____ **Goal 2**: Use slope to solve real-life problems. (p. 78)

Material Covered:

____ Student Help: Skills Review

____ Example 5: Geometrical Use of Slope

____ Example 6: Slope as Rate of Change

____ Other (specify) _____

_____

Homework and Additional Learning Support

____ Textbook (specify) pp. 79–81 _____

_____

____ Internet: Extra Examples at www.mcdougallitell.com

____ *Reteaching with Practice* worksheet (specify exercises) _____

____ *Personal Student Tutor* for Lesson 2.2

NAME _____ DATE _____

Cooperative Learning Activity

For use with pages 75–81

GOAL **To compare temperatures in the Fahrenheit and Celsius scales**

Materials: Graph paper

Background

By constructing a linear graph, the relationship between two variables can be developed.

Instructions

❶ Draw a graph where the *x*-axis represents degrees Celsius and the *y*-axis represents degrees Fahrenheit.

❷ The freezing point of water on each temperature scale is 0°C and 32°F. The boiling point of water on each temperature scale is 32°C and 212°F. Plot these points, (0, 32) and (100, 212), on your graph.

❸ Find the slope of the line using the points.

❹ Find the *y*-intercept.

❺ Using the slope-intercept form of an equation, write an equation that converts a temperature *C* in degrees Celsius to the temperature *F* in degrees Fahrenheit.

Analyzing the Results

1. Does your graph enable you to find any temperature in degrees Fahrenheit, given a temperature in degrees Celsius?

2. Can you find any temperature in degrees Celsius given a temperature in degrees Fahrenheit?

NAME _____ DATE _____

Real–Life Application:
When Will I Ever Use This?

For use with pages 75–81

Economics

The U.S. Department of Commerce (DOC) is the governmental agency in charge of monitoring and making predictions about the U.S. economy. Using the data that they gather from businesses all over the country, the DOC makes predictions about how strong the economy will be in the future.

In many cases, the DOC groups data according to sectors of the economy — small segments of the economy that deal with similar industries. For example, Building Materials, Automotive Sales, Furniture Sales, and General Merchandise are sectors.

One of the variables that the DOC wants to predict is called *national income wage and salary disbursements*—basically, how much money is paid to the employees of the country.

In Exercises 1–8, use the following information.

DOC analysts collected data from four sectors of the economy: Building Materials, Automotive Sales, Furniture Sales, and General Merchandise. The analysts used these data to develop a linear equation to predict national wages based on each one of the sectors. The equations are given in the following table.

Wages vs. Building materials
$W = 551.50 + 0.07635B$
Wages vs. Auto Sales
$W = 603.752 + 0.01895A$
Wages vs. Furniture Sales
$W = 560.92 + 0.08F$
Wages vs. General Merchandise
$W = 923.73 + 0.0246G$

1. Find the slope of each equation.

2. Which line is the steepest?

3. Which of these lines is the least steep?

4. Suppose each of the sectors had an increase of $1 million in sales. Based on the slopes of the equations, (a) which sector would have the most effect on national wages and (b) which would have the least effect?

5. Do you think it is significant that the slopes found in Exercise 4 are positive? Explain.

6. What would it mean if one of these sectors was related to national wages by a negative slope?

7. What would it mean if one of these sectors was related to national wages by a zero slope?

NAME _____ DATE _____

Challenge: Skills and Applications

For use with pages 75–81

In Exercises 1 and 2, use slopes to check whether each set of 3 points lies on one line.

1. $(-1, -3), (2, 1), (8, 9)$

2. $(-4, 7), (0, 2), (5, -2)$

3. Find k so that the line through $(7, 2k)$ and $(4, -3)$ is parallel to the line through $(1, k + 1)$ and $(3, 5)$.

4. Find k so that the line through $(k - 1, k + 2)$ and $(4, -1)$ is perpendicular to the line through $(-3, 2)$ and $(2, 5)$.

5. a. Find the areas of the square and the rectangle shown below.

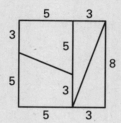 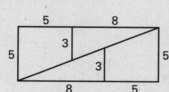

   b. Explain why the two areas you found in part (a) should be the same.

   c. Resolve the apparent contradiction between your results in parts (a) and (b) by using slopes. Draw a diagram to illustrate the actual situation.

6. The curve shown at the right is the graph of $y = x^2$. Line t is a *tangent line* to the graph at A (that is, a line that intersects the graph only at this point and remains on the "same side" of the graph), while line s intersects the graph at points A and B.

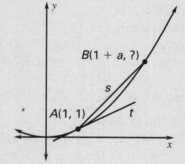

   a. Explain in words what would happen to the relationship between the slopes of lines s and t as the number a approached 0, but point B remained on the graph.

   b. Express the second coordinate of point B in terms of a.

   c. Using your answer to part (b), express the slope of line s in terms of a.

   d. By simplifying your answer to part (c) under the assumption that $a \neq 0$, but imagining a approaching 0, estimate the slope of the line t.

TEACHER'S NAME _____ CLASS _____ ROOM _____ DATE _____

Lesson Plan

1-day lesson (See *Pacing the Chapter,* TE pages 64C–64D)   **For use with pages 82–90**

GOALS 1. **Use the slope-intercept form of a linear equation to graph linear equations.**
2. **Use the standard form of a linear equation to graph linear equations.**

State/Local Objectives _____

_____

✓ **Check the items you wish to use for this lesson.**

STARTING OPTIONS
____ Homework Check: TE page 79; Answer Transparencies
____ Warm-Up or Daily Homework Quiz: TE pages 82 and 81, CRB page 38, or Transparencies

TEACHING OPTIONS
____ Motivating the Lesson: TE page 83
____ Lesson Opener (Application): CRB page 39 or Transparencies
____ Graphing Calculator Activity with Keystrokes: CRB page 40
____ Examples 1–5: SE pages 83–85
____ Extra Examples: TE pages 83–85 or Transparencies
____ Technology Activity: SE page 90
____ Closure Question: TE page 85
____ Guided Practice Exercises: SE page 86

APPLY/HOMEWORK
Homework Assignment
____ Basic 16–18, 20–36 even, 37–40, 44, 46, 52–60 even, 63, 65, 66, 69–85 odd; Quiz 1: 1–11
____ Average 16–18, 20–36 even, 37–40, 44, 46, 52–60 even, 61, 63, 65, 66, 69–85 odd, 86;
         Quiz 1: 1–11
____ Advanced 16–18, 20–36 even, 37–40, 44, 46, 52–60 even, 61, 63–67, 69–85 odd, 86; Quiz 1: 1–11

Reteaching the Lesson
____ Practice Masters: CRB pages 41–43 (Level A, Level B, Level C)
____ Reteaching with Practice: CRB pages 44–45 or Practice Workbook with Examples
____ Personal Student Tutor

Extending the Lesson
____ Applications (Interdisciplinary): CRB page 47
____ Math & History: SE page 89; CRB page 48; Internet
____ Challenge: SE page 88; CRB page 49 or Internet

ASSESSMENT OPTIONS
____ Checkpoint Exercises: TE pages 83–85 or Transparencies
____ Daily Homework Quiz (2.3): TE page 88, CRB page 53, or Transparencies
____ Standardized Test Practice: SE page 88; TE page 88; STP Workbook; Transparencies
____ Quiz (2.1–2.3): SE page 89; CRB page 50

Notes _____

TEACHER'S NAME _____ CLASS _____ ROOM _____ DATE _____

Lesson Plan for Block Scheduling

1-day lesson (See *Pacing the Chapter*, TE pages 64C–64D)   **For use with pages 82–90**

GOALS   1. Use the slope-intercept form of a linear equation to graph linear equations.
    2. Use the standard form of a linear equation to graph linear equations.

State/Local Objectives _____

_____

_____

CHAPTER PACING GUIDE	
Day	**Lesson**
1	2.1 (all); 2.2 (all)
2	**2.3 (all)**
3	2.4 (all)
4	2.5 (all); 2.6 (all)
5	2.7 (all); 2.8 (all)
6	Review/Assess Ch. 2

✓ Check the items you wish to use for this lesson.

STARTING OPTIONS
_____ Homework Check: TE page 79; Answer Transparencies
_____ Warm-Up or Daily Homework Quiz: TE pages 82 and 81, CRB page 38, or Transparencies

TEACHING OPTIONS
_____ Motivating the Lesson: TE page 83
_____ Lesson Opener (Application): CRB page 39 or Transparencies
_____ Graphing Calculator Activity with Keystrokes: CRB page 40
_____ Examples 1–5: SE pages 83–85
_____ Extra Examples: TE pages 83–85 or Transparencies
_____ Technology Activity: SE page 90
_____ Closure Question: TE page 85
_____ Guided Practice Exercises: SE page 86

APPLY/HOMEWORK
Homework Assignment
_____ Block Schedule: 16–18, 20–36 even, 37–40, 44, 46, 52–60 even, 61, 63–67, 69–85 odd, 86; Quiz 1: 1–11

Reteaching the Lesson
_____ Practice Masters: CRB pages 41–43 (Level A, Level B, Level C)
_____ Reteaching with Practice: CRB pages 44–45 or Practice Workbook with Examples
_____ Personal Student Tutor

Extending the Lesson
_____ Applications (Interdisciplinary): CRB page 47
_____ Math & History: SE page 89; CRB page 48; Internet
_____ Challenge: SE page 88; CRB page 49 or Internet

ASSESSMENT OPTIONS
_____ Checkpoint Exercises: TE pages 83–85 or Transparencies
_____ Daily Homework Quiz (2.3): TE page 88, CRB page 53, or Transparencies
_____ Standardized Test Practice: SE page 88; TE page 88; STP Workbook; Transparencies
_____ Quiz (2.1–2.3): SE page 89; CRB page 50

Notes _____

_____

_____

NAME _____ DATE _____

WARM-UP EXERCISES

For use before Lesson 2.3, pages 82–90

Solve for *y* when *x* = 0.

1. $2x + 2y = 12$

2. $y - 4x = 8$

3. $200y + 400x = 1200$

Solve for *y*.

4. $2x + y = 150$

5. $8x - 3y = 6$

DAILY HOMEWORK QUIZ

For use after Lesson 2.2, pages 75–81

Find the slope of the line through the given points. Tell whether the line *rises, falls, is horizontal,* or *is vertical*.

1. $(2, -4), (1, 5)$ **2.** $(3, 4), (-5, 4)$

Tell which line is steeper.

3. Line 1: through $(-3, 5)$ and $(2, -5)$
Line 2: through $(4, -1)$ and $(0, 3)$

4. Line 1: through $(2, 4)$ and $(-1, 7)$
Line 2: through $(3, -1)$ and $(4, 1)$

5. Tell whether the lines are *parallel, perpendicular,* or *neither.*
Line 1: through $(-2, 6)$ and $(2, -4)$
Line 2: through $(-2, -1)$ and $(3, 1)$

6. Find the average rate of change in *y* where *x* is in seconds and *y* is in feet for the points $(5, 6)$ and $(3, 2)$.

**During last night's basketball game, Lisa Schmidt made
30 points in field goals. A field goal is worth 2 or 3 points.**

1. If Lisa made 9 field goals worth 2 points each, how many did
she make worth 3 points each?

2. Let *x* represent the number of 2-point field goals Lisa made and
y represent the number of 3-point field goals she made. Write an
equation to model the situation.

3. Plot several points to graph the equation.

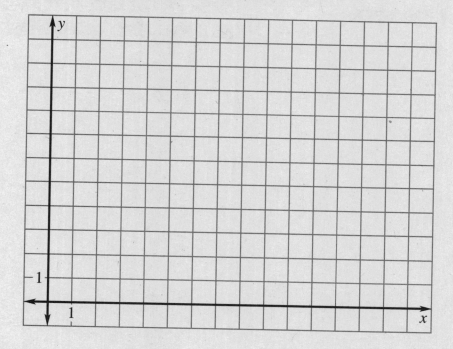

4. Where does the line cross the *x*-axis? What does this point
represent in terms of Lisa's points?

5. Where does the line cross the *y*-axis? What does this point
represent in terms of Lisa's points?

Graphing Calculator Activity Keystrokes

For use with page 90

TI-82

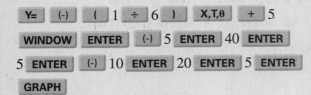

Y= (-) (1 ÷ 6) X,T,θ + 5
WINDOW ENTER (-) 5 ENTER 40 ENTER
5 ENTER (-) 10 ENTER 20 ENTER 5 ENTER
GRAPH

TI-83

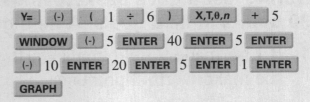

Y= (-) (1 ÷ 6) X,T,θ,n + 5
WINDOW (-) 5 ENTER 40 ENTER 5 ENTER
(-) 10 ENTER 20 ENTER 5 ENTER 1 ENTER
GRAPH

SHARP EL-9600c

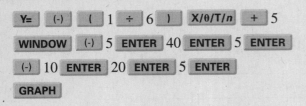

Y= (-) (1 ÷ 6) X/θ/T/n + 5
WINDOW (-) 5 ENTER 40 ENTER 5 ENTER
(-) 10 ENTER 20 ENTER 5 ENTER
GRAPH

CASIO CFX-9850GA PLUS

From the main menu, select GRAPH.

(-) (1 ÷ 6) X,θ,T + 5 EXE
SHIFT F3 (-) 5 EXE 40 EXE 5 EXE
(-) 10 EXE 20 EXE 5 EXE EXIT F6

Practice A

For use with pages 82–89

Draw the line with the given slope and y-intercept.

1. $m = 2, b = 3$

2. $m = 3, b = -1$

3. $m = -1, b = 4$

4. $m = -2, b = -1$

5. $m = 0, b = 6$

6. $m = \frac{1}{3}, b = 2$

7. $m = -\frac{2}{3}, b = 3$

8. $m = \frac{4}{5}, b = -4$

9. $m = -\frac{3}{2}, b = 0$

Find the slope and y-intercept of the line.

10. $y = 3x + 1$

11. $y = -4x + 7$

12. $y = 6x - 4$

13. $y = -8x - 2$

14. $y = \frac{5}{3}x + 1$

15. $y = -\frac{1}{5}x - 3$

Draw the line with the given intercepts.

16. x-intercept: 4
   y-intercept: -2

17. x-intercept: -3
   y-intercept: 1

18. x-intercept: -2
   y-intercept: -4

19. x-intercept: 5
   y-intercept: 2

20. x-intercept: $\frac{1}{2}$

   y-intercept: -3

21. x-intercept: 2

   y-intercept: $-\frac{3}{2}$

Graph the equation.

22. $y = 2x + 1$

23. $y = -6x + 4$

24. $y = 3x - 1$

25. $y = -x + 1$

26. $y = 3x$

27. $y = -2x$

28. *Temperature*   The formula which converts degrees Celsius to degrees Fahrenheit is given by $F = \frac{9}{5}C + 32$. Graph the equation.

Simple Interest   **In Exercises 29–31, use the following information.**
If you deposit \$100 into an account that pays 3% simple interest, the amount of money in your account after t years is modeled by $y = 3t + 100$.

29. What is the slope of the line?

30. What is the y-intercept of the line?

31. Graph the line.

Practice B

For use with pages 82–89

Find the slope and *y*-intercept of the line.

1. $y = 8x - 7$

2. $y = -10x$

3. $x + 4y - 6 = 0$

4. $2x + 4y - 1 = 0$

5. $3x - 7y + 5 = 0$

6. $-2x + 3y - 6 = 0$

Find the intercepts of the line.

7. $y = 3x - 1$

8. $y = -x + 6$

9. $y = \dfrac{2}{3}x + 2$

10. $y = -\dfrac{1}{4}x + 3$

11. $y = \dfrac{5}{3}x - 4$

12. $y = -\dfrac{7}{2}x - 3$

13. $2x - y - 4 = 0$

14. $-3x + 4y - 12 = 0$

15. $5x + 2y + 8 = 0$

16. $x - 3y = 4$

17. $2x + 5y = -8$

18. $-6x + y = 3$

Graph the equation.

19. $y = 4x + 3$

20. $y = -3x - 2$

21. $x + 6y - 3 = 0$

22. $7x - 2y + 6 = 0$

23. $-4x + 8y - 20 = 0$

24. $-6x + 9y = 18$

25. $2x - y = 2$

26. $8x - 2y = 6$

27. $3x - 5y + 15 = 0$

Teeter-Totter   **In Exercises 28–30, use the following information.**
The center post on a teeter-totter is 2 feet high. When one end of the teeter-totter rests on the ground, that end is 7 feet from the center post.

28. Find the slope of the teeter-totter.

29. Assume the base of the center post is at $(0, 0)$ with the ground along the *x*-axis. Find the *y*-intercept of the teeter-totter.

30. Write an equation of the line that follows the path of the teeter-totter.

31. *Saving Change*   Each time you get dimes or quarters for change, you throw them into a jar. You are hoping to save $50. Write a model that shows the different numbers of dimes and quarters that you could accumulate to reach your goal.

32. *Commission Sales*   A salesperson receives a 3% commission on furniture sold at a sale price and a 4% commission on furniture sold at the regular price. The salesperson wants to earn a $250 commission. Write a model that shows the different amounts of sale-priced and regular-priced furniture that can be sold to reach this goal.

NAME _____ DATE _____

Practice C

For use with pages 82–89

Find the slope and the y-intercept of the line.

1. $y = 4x + 2$

2. $y = -3x + \dfrac{1}{2}$

3. $y = -\dfrac{2}{3}x + 4$

4. $y = -3 + 2x$

5. $y = 6$

6. $4x - 3y + 1 = 0$

7. $7x + 5y - 8 = 0$

8. $-3x + 2y + 4 = 0$

9. $-8x + 3y = 0$

10. $-2x + 5y - 7 = 0$

11. $3x - 7y + 1 = 0$

12. $x + 2y - 5 = 0$

Find the intercepts of the line.

13. $3x + 4y - 12 = 0$

14. $2x - y + 8 = 0$

15. $3x + 2y - 5 = 0$

16. $5x - 2y = 0$

17. $4x + y = 3$

18. $x + 13 = 0$

19. $4y - 3 = 0$

20. $2x + 3y = 3x - y + 1$

21. $x - 5y + 3 = 3x - y + 4$

Graph the equation.

22. $y = -3x + 5$

23. $y = -2x - \dfrac{1}{2}$

24. $y = \dfrac{3}{4}x + 1$

25. $x = \dfrac{4}{3}$

26. $2x + 3y + 6 = 0$

27. $3x - 4y = 10$

28. $-x + 2y - 8 = 0$

29. $\dfrac{1}{2}x + 2y - 3 = 0$

30. $4x - \dfrac{3}{2}y - 1 = 0$

31. *Fund Raiser* The marching band holds a fund raiser each year in which they sell t-shirts and sweatshirts with the school's name and mascot on it. The t-shirts sell for $7 and the sweatshirts sell for $15. The band needs to raise $3000. Write a model that shows the number of t-shirts and sweatshirts that must be sold. Then graph the model and determine three combinations of t-shirts and sweatshirts that satisfy the model.

32. *Linear Depreciation* A business purchases a piece of equipment for $300,000. The value, V, of the machine after t years is represented by the model $2V + 100,001t = 600,000$.

a. Find the V-intercept of the model. What does the V-intercept represent?

b. Find the slope of the model. What does the slope represent?

Lesson 2.3

Reteaching with Practice

For use with pages 82–89

GOAL Use the slope-intercept and standard forms of linear equations to graph linear equations

VOCABULARY

The **slope-intercept form** of a linear equation is $y = mx + b$, where m is the slope and b is the y-intercept.

The **standard form** of a linear equation is $Ax + By = C$, where A and B are not both zero.

The **y-intercept** is the y-coordinate of the point where the graph crosses the y-axis and is found by letting $x = 0$ and solving for y.

The **x-intercept** is the x-coordinate of the point where the graph crosses the x-axis and is found by letting $y = 0$ and solving for x.

EXAMPLE 1 *Graphing with the Slope-Intercept Form*

Graph $2x + y = 3$.

SOLUTION

1. Write the equation in slope-intercept form by subtracting $2x$ from each side.

$$y = -2x + 3$$

2. The y-intercept is 3, so plot the point $(0, 3)$.

3. The slope is $-\frac{2}{1}$, so plot a second point by moving 1 unit to the left and 2 units up. This point is $(-1, 5)$.

4. Draw a line through the two points.

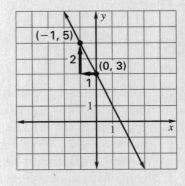

Exercises for Example 1

Graph the equation.

1. $y = 3x - 1$ **2.** $y = -\frac{2}{3}x + 2$ **3.** $-3x - y = 4$

EXAMPLE 2 *Graphing with the Standard Form*

Graph $-2x + 3y = -6$.

SOLUTION

1. The equation is already in standard form.

2. $-2x + 3(0) = -6$ Let $y = 0$.

           $x = 3$ Solve for x.

The x-intercept is 3, so plot $(3, 0)$.

continued

Reteaching with Practice

For use with pages 82–89

3. $-2(0) + 3y = -6$ Let $x = 0$.

   $y = -2$ Solve for y.

   The y-intercept is -2, so plot $(0, -2)$.

4. Draw a line through the two points.

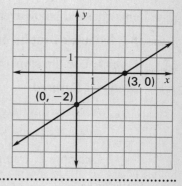

Exercises for Example 2

Graph the equation.

4. $-x + 4y = 8$

5. $2x + 5y = -10$

6. $2x - y = 4$

7. $x = -3$

8. $y = -6$

9. $x = 4$

EXAMPLE 3   *Using the Standard Form*

Sales for the firefighter's benefit dinner were $1980. The cost for a child's dinner was $4.50 and an adult's dinner was $6.00. Describe the number of children and adults who attended to reach this amount.

SOLUTION

Verbal Model	Cost per child	·	Number of children	+	Cost per adult	·	Number of adults	=	Total revenue

Algebraic Model $4.50c + 6.00a = 1980$

The graph of $4.5c + 6a = 1980$ is a line that inter-sects the c-axis at $(440, 0)$ and the a-axis at $(0, 330)$. Points with integer coefficients on the line segment joining $(440, 0)$ and $(0, 330)$ represent the numbers of children and adults that could have attended. One way to earn $1980 would be to sell 200 children's tickets and 180 adult's tickets.

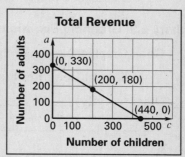

Exercises for Example 3

10. Members of the chorus need to raise $650 selling hoagies and pizzas. The profit they receive on each hoagie is $1.25 and $2.50 for each pizza. Find three different combinations for the numbers of each sold to raise the needed amount.

11. Daniel is employed at two different restaurants. At one restaurant he makes $5.35 per hour and at the other he makes $6.25. Daniel needs to earn $105.30 a week to pay his bills. Describe the number of hours he could work at each restaurant to earn this amount.

NAME _____ DATE _____

Quick Catch-Up for Absent Students

For use with pages 82–90

The items checked below were covered in class on (date missed) _____

Lesson 2.3: Quick Graphs of Linear Equations

_____ **Goal 1:** Use the slope-intercept form of a linear equation to graph linear equations. (pp. 82–83)

Material Covered:

_____ Activity: Investigating Slope and *y*-intercept

_____ Example 1: Graphing with the Slope-Intercept Form

_____ Example 2: Using the Slope-Intercept Form

Vocabulary:

    y-intercept, p. 84   slope-intercept form of a linear equation, p. 84

_____ **Goal 2:** Use the standard form of a linear equation to graph linear equations. (pp. 84–85)

Material Covered:

_____ Student Help: Look back

_____ Example 3: Drawing Quick Graphs

_____ Example 4: Graphing Horizontal and Vertical Lines

_____ Student Help: Study Tip

_____ Example 5: Using the Standard Form

Vocabulary:

    standard form of a linear equation, p. 84

    x-intercept, p. 84

Activity 2.3: Graphing Equations (p. 90)

_____ **Goal:** Graph equations of the form $y = f(x)$ using a graphing calculator.

_____ Student Help: Keystroke Help

_____ Other (specify) _____

_____

Homework and Additional Learning Support

_____ Textbook (specify) <u>pp. 86–89</u> _____

_____

_____ *Reteaching with Practice* worksheet (specify exercises) _____

_____ *Personal Student Tutor* for Lesson 2.3

Lesson 2.3

Interdisciplinary Application

For use with pages 82–89

The Internet

COMPUTER SCIENCE   The Internet has become a major tool for any institution that uses computers. Indeed, some businesses do nothing but work in designing material to be exchanged over the Internet through the World Wide Web (WWW). To transmit information on the WWW, designers and programmers must translate the information into a special computer language known as HTML (Hypertext Markup Language), thus making it available for others to view.

One of the major concerns of these WWW designers is the concept of load time—the time it takes for a computer to transfer an entire document from one computer to another. There are many factors that influence this speed-of-transfer, but the most obvious is the length of the document. Longer documents take longer to transfer.

In Exercises 1–6, use the following information.

Suppose you were starting a company that wanted to provide complete works of literature over the Internet for English classes to use. The following questions have to do with some decisions you will have to make in setting up your business.

The HTML programmers gathered some sample novels and calculated that, if x is the number of words in a novel and y is load time of the novel (in seconds), then the relationship between the two is

$$y = 0.003x + 2.2.$$

1. What is the slope of the line given by this equation?

2. Using your own words, say what the slope represents.

3. What is the y-intercept of this equation?

4. Using your own words, say what the y-intercept represents.

5. Write the equation in standard form.

6. Graph the equation.

In Exercises 7–11, calculate the load times of the following works of literature using the given word counts.

7. *Gulliver's Travels* by Jonathan Swift: 106,014 words

8. *Flatland* by Edwin Abbot: 14,784 words

9. *Titus Andronicus* by William Shakespeare: 22,962 words

10. Research has determined that users will not want to wait more than 60 seconds for a document to load. Which of the works in Exercises 7, 8, and 9 will load in this time?

11. What would you suggest as a remedy for pages that take too long to load?

Math and History Application

For use with page 89

HISTORY   The race to set transatlantic speed records has a long history. In July of 1845 the clipper ship *James Baines* set a record for sailing ships by sailing from Boston to Liverpool in 12 days and 6 hours. Just a few years ago, a swimmer named Ben Lecomte set a much slower but equally amazing record. It took Lecomte 80 days to swim from Hyannis, MA, to Quiberon on the French coast with the aid of a monofin attached to both feet. Ben was accompanied by a ship, and swam inside a "protective ocean device" that used electric fields to repel sharks.

Migrating (and lost) birds have also set some impressive records. The current distance record holder appears to be a Common Tern. Banded as a chick in Finland on June 30, 1996, it was caught on January 24, 1997 on a beach in southeastern Australia! Scientists estimate that it flew about 26,000 kilometers in at most 208 days, for a minimum average speed of 3.2 miles per hour.

In the Math and History feature on page 89 you may have noticed that the QE2's speed was given in knots or nautical miles per hour. The nautical mile is about 1.151 statute or land miles, which seems like a strange number. This "sea mile," which has been used by mariners since the seventeenth century, is based on the practice of measuring latitude and longitude in degrees. If you traveled all the way around the earth at the equator, you'd cover 360 degrees, and each degree is divided into 60 minutes. The nautical mile is defined so that one nautical mile equals one minute along any great circle. Aviators also navigate with degrees and minutes, so airplane speeds are also usually measured in knots.

MATH   Here are some problems about transatlantic travel by ships, people, and birds.

1. If the Titanic had missed the iceberg and kept on going at the average speed that you computed in the Math and History feature, when would it have reached New York?

2. The distance from Boston to Liverpool is about 3150 miles. How many times faster was the Titanic's average speed traveling from Cobh to the iceberg than the average speed of the *James Baines* on its record-setting trip?

3. Find Ben Lecomte's approximate speed in knots for his transatlantic swim, which covered 3736 nautical miles.

4. When not swimming, Ben rested on the ship, which was allowed to drift with currents and winds, so that his progress would be due solely to his swimming. What do you think a graph of his distance from Hyannis plotted against time would look like? Would it be linear? Would its slope always be positive?

NAME _____ DATE _____

Challenge: Skills and Applications

For use with pages 82–89

1. A line passes through the points $(-1, 5)$ and $(3, k)$ and has y-intercept 7. Find the value of k.

2. A line with x-intercept -3 and y-intercept k passes through the point $(5, 8)$. Find the value of k.

3. A line has x-intercept p and y-intercept q. Express the equation of the line in terms of p and q.

4. Suppose you know that a line whose equation is $y = mx + b$ has the property that for two specific (distant) points on the line (x_1, y_1) and (x_2, y_2),

$$\frac{x_1}{x_2} = \frac{y_1}{y_2}.$$

   Show that b must be 0.

5. A line can be defined by a pair of *parametric equations* in a new variable t.

$$x = pt + q$$
$$y = rt + s$$

   Thinking of t as representing time, you can imagine a particle at position (x, y) at time t tracing out the line.

   a. By solving the first equation above for t and substituting the expression you get into the second equation, write the two equations as one linear equation in slope-intercept form. What is the slope of the line, in terms of the constants p, q, r, and s? What is its y-intercept?

   b. Show that this pair of equations

$$x = q - pt$$
$$y = s - rt$$

   gives the same slope-intercept form. Explain this in terms of particle motion.

6. Suppose the line with equation $y = mx + b$ is translated 3 units to the right. Write the new equation in slope-intercept form. What is the new y-intercept?

7. Suppose the line in Exercise 6 is translated 2 units to the left and 5 units up. Write the new equation in slope-intercept form.

NAME _____ DATE _____

Quiz 1

For use after Lessons 2.1–2.3

Answers

1. Identify the domain and range. Then tell whether the relation is a function. *(Lesson 2.1)*

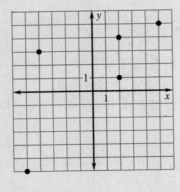

1. _____

   _____

   _____

2. _____

3. _____

4. _____

5. _____

6. Use grid at left. _____

7. Use grid at left. _____

2. Decide whether the function is linear. *(Lesson 2.1)*

   $y = x^2 + 2x - 2$

3. Find the indicated value of $f(x)$. *(Lesson 2.1)*

   $f(x) = 3x^2 - x + 8; \ f(-3)$

4. Tell whether the lines are *parallel, perpendicular,* or *neither.* *(Lesson 2.2)*

   Line 1: through $(3, 8)$ and $(7, 4)$

   Line 2: through $(-2, 5)$ and $(-5, 8)$

5. Tell whether the lines are *parallel, perpendicular,* or *neither.* *(Lesson 2.2)*

   Line 1: through $(5, -2)$ and $(6, 10)$

   Line 2: through $(7, -4)$ and $(19, -5)$

6. Graph the equation. *(Lesson 2.3)*

   $y = -2x + 4$

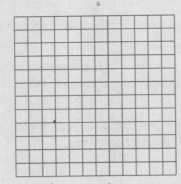

7. Graph the equation. *(Lesson 2.3)*

   $y = -8$

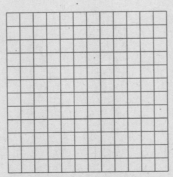

TEACHER'S NAME _____ CLASS _____ ROOM _____ DATE _____

Lesson Plan

2-day lesson (See *Pacing the Chapter,* TE pages 64C–64D)   **For use with pages 91–98**

GOALS   1. **Write linear equations.**
2. **Write direct variation equations.**

State/Local Objectives _____

_____

✓ **Check the items you wish to use for this lesson.**

STARTING OPTIONS

_____ Homework Check: TE page 86; Answer Transparencies
_____ Warm-Up or Daily Homework Quiz: TE pages 91 and 88, CRB page 53, or Transparencies

TEACHING OPTIONS

_____ Motivating the Lesson: TE page 92
_____ Lesson Opener (Application): CRB page 54 or Transparencies
_____ Examples: Day 1: 1–4, SE pages 91–93; Day 2: 5–7, SE pages 93–94
_____ Extra Examples: Day 1: TE pages 92–93 or Transp.; Day 2: TE pages 93–94 or Transp.; Internet
_____ Closure Question: TE page 94
_____ Guided Practice: SE page 95 Day 1: Exs. 2, 4–11; Day 2: Exs. 1, 3, 12

APPLY/HOMEWORK
Homework Assignment

_____ Basic Day 1: 14–22 even, 29–41 odd, 42; Day 2: 25, 26–40 even, 43–59 odd, 69, 71–93 odd
_____ Average Day 1: 14–28 even, 25–51 odd; Day 2: 44–58 even, 59–65 odd, 69, 71–93 odd
_____ Advanced Day 1: 14–28 even, 25–57 odd, 60, 62; Day 2: 56, 58, 59–69 odd, 70, 71–93 odd

Reteaching the Lesson

_____ Practice Masters: CRB pages 55–57 (Level A, Level B, Level C)
_____ Reteaching with Practice: CRB pages 58–59 or Practice Workbook with Examples
_____ Personal Student Tutor

Extending the Lesson

_____ Applications (Real-Life): CRB page 61
_____ Challenge: SE page 98; CRB page 62 or Internet

ASSESSMENT OPTIONS

_____ Checkpoint Exercises: Day 1: TE pages 92–93 or Transp.; Day 2: TE pages 93–94 or Transp.
_____ Daily Homework Quiz (2.4): TE page 98, CRB page 65, or Transparencies
_____ Standardized Test Practice: SE page 98; TE page 98; STP Workbook; Transparencies

Notes _____

_____

_____

TEACHER'S NAME _____ CLASS _____ ROOM _____ DATE _____

Lesson Plan for Block Scheduling

1-day lesson (See *Pacing the Chapter,* TE pages 64C–64D)   **For use with pages 91–98**

 GOALS 1. **Write linear equations.**
2. **Write direct variation equations.**

State/Local Objectives _____

_____

✓ **Check the items you wish to use for this lesson.**

STARTING OPTIONS

_____ Homework Check: TE page 86; Answer Transparencies
_____ Warm-Up or Daily Homework Quiz: TE pages 91 and 88,
           CRB page 53, or Transparencies

TEACHING OPTIONS

_____ Motivating the Lesson: TE page 92
_____ Lesson Opener (Application): CRB page 54 or Transparencies
_____ Examples: 1–7, SE pages 91–94;
_____ Extra Examples: TE pages 92–94 or Transparancies.; Internet
_____ Closure Question: TE page 94
_____ Guided Practice Exercises: SE page 95

APPLY/HOMEWORK

Homework Assignment
_____ Block Schedule: 14–24 even, 25–61 odd, 67, 69, 71–93 odd

Reteaching the Lesson
_____ Practice Masters: CRB pages 55–57 (Level A, Level B, Level C)
_____ Reteaching with Practice: CRB pages 58–59 or Practice Workbook with Examples
_____ Personal Student Tutor

Extending the Lesson
_____ Applications (Real Life): CRB page 61
_____ Challenge: SE page 98; CRB page 62 or Internet

ASSESSMENT OPTIONS

_____ Checkpoint Exercises: TE pages 92–94 or Transparancies
_____ Daily Homework Quiz (2.4): TE page 98, CRB page 65, or Transparencies
_____ Standardized Test Practice: SE page 98; TE page 98; STP Workbook; Transparencies

Notes _____

_____

_____

CHAPTER PACING GUIDE	
Day	**Lesson**
1	2.1 (all); 2.2 (all)
2	2.3 (all)
3	**2.4 (all)**
4	2.5 (all); 2.6 (all)
5	2.7 (all); 2.8 (all)
6	Review/Assess Ch. 2

Algebra 2
Chapter 2 Resource Book

LESSON
2.4
NAME _____ DATE _____

WARM-UP EXERCISES

For use before Lesson 2.4, pages 91–98

1. Find the slope and the y-intercept of $y = 2x - 4$.

2. Find the slope and the y-intercept of $y = \dfrac{3}{2}x + 1$.

3. Find the slope of a line parallel to the line in Exercise 1.

4. Find the slope of a line perpendicular to the line in Exercise 2.

DAILY HOMEWORK QUIZ

For use after Lesson 2.3, pages 82–90

1. Draw the line with slope $m = -2$ and y-intercept $b = -1$.

2. Graph $y = \dfrac{1}{2}x - 2$.

3. Find the slope and the y-intercept of the line $-3x + y = -8$.

4. Graph $3x - 2y = 6$. Label any intercepts.

Algebra 2
Chapter 2 Resource Book

53

Lesson 2.4

Available as a transparency

Application Lesson Opener

For use with pages 91–98

The volume of blood your heart pumps is related to your pulse rate. The table shows the volume of blood pumped for several pulse rates.

Pulse rate (p) (in beats per minute)	60	72	85	94
Volume of blood pumped (V) (in liters)	3.6	4.32	5.1	5.64
$\dfrac{V}{p}$				

1. Divide the volume by the related pulse rate to complete the table.

2. Write an equation of the form $V = kp$ that represents the relationship in the table.

The table shows the number of minutes of exercise it takes to burn 150 calories for several activities.

Activity	Bicycling	Walking	Swimming laps	Washing windows
Minutes	15	30	20	50
$\dfrac{\text{Cal}}{\text{min}}$				

3. Divide to find the number of calories burned per minute.

4. Let c represent the number of calories burned and let t represent the number of minutes spent exercising. Is there a single value you can use for k to write an equation of the form $c = kt$ that will work for all activities? Explain.

Lesson 2.4

NAME _____ DATE _____

Practice A

For use with pages 91–98

Write an equation of the line that has the given slope and y-intercept.

1. $m = 3, b = 2$ **2.** $m = 4, b = -5$ **3.** $m = -6, b = 1$

4. $m = -1, b = -9$ **5.** $m = 2, b = 0$ **6.** $m = 0, b = 7$

Write an equation of the line that passes through the given point and has the given slope.

7. $(0, 3), m = 5$ **8.** $(0, -2), m = 3$ **9.** $(1, 2), m = -2$

10. $(3, 1), m = 4$ **11.** $(-2, 6), m = 0$ **12.** $(4, -1), m = 1$

13. $(5, -2), m = -1$ **14.** $(-3, -7), m = 2$ **15.** $(8, 2), m = -1$

Write an equation of the line that passes through the given points.

16. $(1, 1), (5, 9)$ **17.** $(2, 1), (3, -7)$ **18.** $(-1, 4), (2, 16)$

19. $(-3, -2), (-1, 0)$ **20.** $(5, 5), (8, -4)$ **21.** $(0, 3), (4, 0)$

22. $(2, -1), (1, -6)$ **23.** $(7, 8), (2, 18)$ **24.** $(9, 4), (1, 8)$

Write an equation of the line.

25. 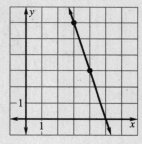   **26.** 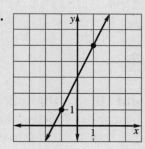   **27.**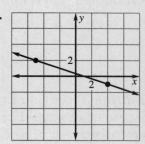

Tell whether the data show direct variation. If so, write an equation relating x and y.

28.

x	1	2	3	4	5
y	2	3	4	5	6

29.

x	-2	-1	0	1	2
y	2	1	0	-1	-2

30. *Sales Tax*   The amount of sales tax in Pennsylvania varies directly with the price of merchandise. Use the given tax table to write an equation relating the price x and the amount of sales tax y.

Price, x (dollars)	10	20	30	40	50
Tax, y (dollars)	0.60	1.20	1.80	2.40	3

Practice B

For use with pages 91–98

Write an equation of the line that has the given slope and *y*-intercept.

1. $m = 4, b = -4$

2. $m = -6, b = 3$

3. $m = \dfrac{4}{3}, b = 6$

4. $m = -\dfrac{1}{2}, b = -4$

5. $m = 8, b = 0$

6. $m = 0, b = 5$

Write an equation of the line that passes through the given point and has the given slope.

7. $(2, 1), m = -2$

8. $(-4, 3), m = 5$

9. $(7, -5), m = 1$

10. $(-1, -10), m = 3$

11. $\left(\dfrac{1}{2}, 4\right), m = -8$

12. $\left(\dfrac{2}{3}, 0\right), m = -4$

Write an equation of the line that passes through the given points.

13. $(-2, 1), (2, 4)$

14. $(-1, 3), (1, -1)$

15. $(-3, -1), (3, 2)$

16. $(4, -2), (6, -3)$

17. $(1, 5), (-4, 0)$

18. $(3, -7), (-2, 3)$

19. $(-6, 1), (-5, 4)$

20. $(-3, -2), (4, 1)$

21. $(10, -4), (6, -10)$

The variables *x* and *y* vary directly. Write an equation that relates the variables. Then find *y* when *x* = 10.

22. $x = 2, y = 6$

23. $x = -1, y = 5$

24. $x = 4, y = -10$

25. $x = 1, y = 0.25$

26. $x = -8, y = 2$

27. $x = \dfrac{1}{3}, y = \dfrac{9}{10}$

Measuring Speed **In Exercises 28 and 29, use the following information.**

The speed of an automobile in miles per hour varies directly with its speed in kilometers per hour. A speed of 64 miles per hour is equivalent to a speed of 103 kilometers per hour.

28. Write a linear model that relates speed in miles per hour to speed in kilometers per hour.

29. You are driving through Canada and see a speed limit sign that says the speed limit is 80 kilometers per hour. You are traveling 55 miles per hour. Are you speeding?

Fish and Shellfish Consumption **In Exercises 30 and 31, use the following information.**

For 1992 through 1994, the per capita consumption of fish and shellfish in the U.S. increased at a rate that was approximately linear. In 1992, the per capita consumption was 14.7 pounds, and in 1994 it was 15.1 pounds.

30. Write a linear model for the per capita consumption of fish and shellfish in the U.S. Let *t* represent the number of years since 1992.

31. What would you expect the per capita consumption of fish and shellfish to be in 2002?

NAME _____ DATE _____

Practice C

For use with pages 91–98

Write an equation of the line that passes through the given points.

1. $(2, -1), (3, 8)$ **2.** $(-5, 3), (2, 2)$ **3.** $(1, -6), (1, -2)$

4. $(7, 2), (-4, -6)$ **5.** $(3, 8), (-1, 8)$ **6.** $(6, 6), (-2, -2)$

**Write an equation of the line that passes through the given point
and is perpendicular to the given line.**

7. $(1, 3), \ y = 2x - 1$ **8.** $(-3, 2), \ y = -4x + 3$ **9.** $(1, 1), \ y = \frac{1}{2}x - 7$

10. $(-3, 1), \ y = -\frac{2}{3}x + 4$ **11.** $(7, -3), \ y = 8$ **12.** $(5, 2), \ x = 2$

**Write an equation of the line that passes through the given point
and is parallel to the given line.**

13. $(-2, 1), \ y = 2x + 5$ **14.** $(1, -1), \ y = -x + 3$ **15.** $(-3, -5), \ y = 12 + x$

16. $(3, -4), \ y = \frac{1}{2}x - 8$ **17.** $(10, -12), \ y = -\frac{3}{4}x + 1$ **18.** $(4, -9), \ y = 14$

Labor Force   **In Exercises 19–21, use the following information.**

From 1840 to 1850, the rate at which the percent of the labor force in nonfarming occupations increased was approximately linear. In 1840, 31.4% of the labor
force held nonfarming jobs. In 1850, 36.3% of the labor force held nonfarming
jobs.

19. Write a linear model for the percentage of the labor force in nonfarming
occupations. Let $t = 0$ represent 1840.

20. In 1860, the percent of the labor force in nonfarming occupations was
41.1%. Is the model for the percentage of nonfarming occupations from
1840 to 1850 still an appropriate model?

21. In 1870, the percent of the labor force in nonfarming occupations was
47.0%. Is the model for the percentage of nonfarming occupations from
1840 to 1850 still an appropriate model?

College Tuition   **In Exercises 22–24, use the following information.**

The rate of increase in tuition at a college from 1990 to 1995 was approximately
linear. In 1990, the tuition was $15,500 and in 1995 it was $22,600.

22. Write a linear model for the tuition from 1990 to 1995. Let $t = 0$
represent 1990.

23. Write a linear model for the tuition from 1990 to 1995. Use the actual
years as the coordinates for time.

24. Although the models in Exercises 22 and 23 are different, use both
models to approximate the tuition in 2000. Do both models yield the
same result?

NAME _____ DATE _____

Reteaching with Practice

For use with pages 91–98

GOAL Write linear equations and direct variation equations

VOCABULARY

Two variables x and y show **direct variation** provided $y = kx$ and $k \neq 0$.

The nonzero constant k is called the **constant of variation.**

EXAMPLE 1 *Writing an Equation Given the Slope and y-intercept*

Write an equation of the line that has $m = -\frac{2}{3}$ and $b = -2$.

SOLUTION

$y = mx + b$ Use slope-intercept form.

$y = -\frac{2}{3}x - 2$ Substitute $-\frac{2}{3}$ for m and -2 for b.

An equation of the line is $y = -\frac{2}{3}x - 2$.

Exercises for Example 1
....

Write an equation of the line that has the given slope and y-intercept.

1. $m = 3$, $b = 0$ **2.** $m = \frac{3}{4}$, $b = 2$ **3.** $m = -2$, $b = -3$

EXAMPLE 2 *Writing an Equation Given the Slope and a Point*

Write an equation of the line that passes through $(-1, -3)$ and has a slope of 4.

SOLUTION

$y - y_1 = m(x - x_1)$ Use point-slope form.

$y + 3 = 4(x + 1)$ Substitute for m, x_1, and y_1.

$y + 3 = 4x + 4$ Distributive property

$y = 4x + 1$ Write in slope-intercept form.

Exercises for Example 2
....

Write an equation of the line that passes through the given point and has the given slope.

4. $(2, -1)$, $m = -5$ **5.** $(0, 5)$, $m = \frac{1}{3}$ **6.** $(-3, -2)$, $m = 0$

EXAMPLE 3 *Writing an Equation Given Two Points*

Write an equation of the line that passes through $(-1, -3)$ and $(2, 4)$.

NAME _____ DATE _____

Reteaching with Practice

For use with pages 91–98

SOLUTION

First, find the slope by letting $(x_1, y_1) = (-1, -3)$ and $(x_2, y_2) = (2, 4)$.

$$m = \frac{y_2 - y_1}{x_2 - x_1} = \frac{4 - (-3)}{2 - (-1)} = \frac{7}{3}$$

Because you know the slope and a point on the line, use the point-slope form to find an equation of the line.

$y - y_1 = m(x - x_1)$	Use point-slope form.
$y + 1 = \frac{7}{3}(x + 3)$	Substitute for m, x_1, and y_1.
$y + 1 = \frac{7}{3}x + 7$	Distributive property
$y = \frac{7}{3}x + 6$	Write in slope-intercept form.

Exercises for Example 3

Write an equation of the line that passes through the given points.

7. $(2, 5)$ and $(4, -1)$ **8.** $(-2, 1)$ and $(4, 7)$ **9.** $(-5, 0)$ and $(0, -1)$

EXAMPLE 4   *Writing and Using a Direct Variation Equation*

The variables x and y vary directly, and $y = 3$ when $x = -4$. Write an equation that relates the variables. Then find y when $x = 4$.

SOLUTION

$$y = kx$$

$$3 = k(-4)$$

$$-\frac{3}{4} = k$$

The direct variation equation is $y = -\frac{3}{4}x$.

When $x = 4$, the value of y is $y = -\frac{3}{4}(4) = -3$.

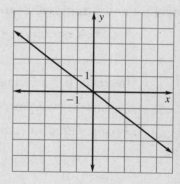

Exercises for Example 4

The variables x and y vary directly. Write an equation that relates the variables. Then find y when $x = -2$.

10. $x = 10$, $y = 100$ **11.** $x = -3$, $y = 12$ **12.** $x = 18$, $y = -2$

NAME _____ DATE _____

Quick Catch-Up for Absent Students

For use with pages 91–98

The items checked below were covered in class on (date missed) _____

Lesson 2.4: Writing Equations of Lines

_____ **Goal 1:** Write linear equations. (pp. 91–93)

Material Covered:

_____ Example 1: Writing an Equation Given the Slope and *y*-intercept

_____ Example 2: Writing an Equation Given the Slope and a Point

_____ Example 3: Writing Equations of Perpendicular and Parallel Lines

_____ Example 4: Writing an Equation Given Two Points

_____ Example 5: Writing and Using a Linear Model

_____ **Goal 2:** Write direct variation equations. (p. 94)

Material Covered:

_____ Example 6: Writing and using a Direct Variation Equation

_____ Example 7: Identifying Direct Variation

Vocabulary:

   direct variation, p. 94   constant of variation, p. 94

_____ Other (specify) _____

   _____

Homework and Additional Learning Support

_____ Textbook (specify) pp. 95–98 _____

   _____

_____ Internet: Extra Examples at www.mcdougallitell.com

_____ *Reteaching with Practice* worksheet (specify exercises) _____

_____ *Personal Student Tutor* for Lesson 2.4

NAME _____ DATE _____

Real–Life Application: When Will I Ever Use This?

For use with pages 91–98

Psychology

Psychometricians are people who specialize in tests and measurements of mental processes, such as achievement tests and personality tests. They frequently use psychological tests to help with career counselling or to select people to receive special training.

Many times, the actual scores from a group of people have to be scaled for interpretation or standardization. This aids the psychometricians in reporting scores to people that have not received the special training needed to interpret the raw scores.

In Exercises 1–8, use the following information.

A psychometrician develops a test to determine the speed at which individuals read. Scores on this test range from 0 (very slow readers) to 22 (very fast readers). However, the psychologist wants the scores to be converted to a scale ranging from 0 to 100.

To convert the scores, the psychometrician wants people who scored 22 to have a "converted" score of 100, while people who scored 0 to maintain their score of 0. All scores in-between will be scaled appropriately.

1. Write two ordered pairs to represent the lowest and highest possible scores on the test. The ordered pairs should be of the form (old score, new score).

2. Find the equation of the line passing through the points in Exercise 1.

3. Use the equation found in Exercise 2 to find the scaled scores of the following people.

Name	Original Score	Scaled Score
Wade	15	
Felecia	18	
Lee	10	
Brandy	8	
Tonya	21	

4. Do you think a scores from 0 to 100 is easier to interpret than a range of scores from 0 to 22?

5. Suppose the psychometrician wanted to re-scale the scores on a scale of 1 to 10, where 1 represented the fastest readers and 10 represented the slowest. What two ordered pairs, of the form (old score, new score), would represent this new situation?

6. Find the equation of the line passing through the points in Exercise 5.

7. Use the equation found in Exercise 6 to re-scale the scores of the following people.

Name	Original Score	Scaled Score
Wade	15	
Felecia	18	
Lee	10	
Brandy	8	
Tonya	21	

NAME _____ DATE _____

Challenge: Skills and Applications

1. Find the value of k that makes the line through $(2, -3)$ and $(5, k)$ have y-intercept 4.

2. Find the value of k that makes the line through $(-4, k)$ and $(-1, 2k)$ have y-intercept -7.

3. Suppose a line goes through the points (x_1, y_1) and (x_2, y_2). Express the y-intercept of the line in terms of these four coordinates.

4. A line can be defined by a set of parametric equations that specify the coordinate of a point (x, y) on the line as functions of a third variable t, often thought of as representing time. Suppose a certain line is defined by the following parametric equations.

$$x = 3(1 - t) - 5t$$
$$y = -2(1 - t) + 2t.$$

   a. Find the coordinates of the 2 points on the line that are associated with $t = 0$ and $t = 1$, and use these to give an equation for the line in point-slope form.

   b. Find the point on the line associated with $t = \frac{1}{2}$. How is this point related *geometrically* to the two points you found in part (a). Make a conjecture about any point associated with a value of t between 0 and 1.

5. Show that if the points (x_1, y_1) and (x_2, y_2) are on the graph of $y = mx + b$, then

$$\frac{y_1 - b}{y_2 - b} = \frac{x_1}{x_2}.$$

   (*Hint:* This equation will be true if and only if $x_2(y_1 - b) = x_1(y_2 - b)$. Show that the two sides of this equation reduce to the same expression.)

6. In the slope-intercept form $y = mx + b$, suppose b is fixed but m is allowed to vary. Describe in words the family of lines that results from using all possible values of m.

7. In the point-slope form $y - y_1 = m(x - x_1)$, describe in words the family of lines that results when m is allowed to vary through all real numbers. Does the special case $m = 0$ fit your description?

TEACHER'S NAME _____ CLASS _____ ROOM _____ DATE _____

Lesson Plan

1-day lesson (See *Pacing the Chapter*, TE pages 64C–64D)   For use with pages 99–107

GOALS 1. **Use a scatter plot to identify the correlation shown by a set of data.**
2. **Approximate the best-fitting line for a set of data.**

State/Local Objectives _____

_____

✓ Check the items you wish to use for this lesson.

STARTING OPTIONS
_____ Homework Check: TE page 95; Answer Transparencies
_____ Warm-Up or Daily Homework Quiz: TE pages 100 and 98, CRB page 65, or Transparencies

TEACHING OPTIONS
_____ Motivating the Lesson: TE page 101
_____ Concept Activity: SE page 99; CRB page 66 (Activity Support Master)
_____ Lesson Opener (Application): CRB page 67 or Transparencies
_____ Graphing Calculator Activity with Keystrokes: CRB page 68
_____ Examples 1–3: SE pages 100–102
_____ Extra Examples: TE pages 101–102 or Transparencies
_____ Technology Activity: SE page 107
_____ Closure Question: TE page 102
_____ Guided Practice Exercises: SE page 103

APPLY/HOMEWORK
Homework Assignment
_____ Basic 8–24 even, 25, 28, 31–43 odd; Quiz 2: 1–9
_____ Average 8–24 even, 25–28, 31–43 odd; Quiz 2: 1–9
_____ Advanced 8–24 even, 25–29, 31–43 odd; Quiz 2: 1–9

Reteaching the Lesson
_____ Practice Masters: CRB pages 69–71 (Level A, Level B, Level C)
_____ Reteaching with Practice: CRB pages 72–73 or Practice Workbook with Examples
_____ Personal Student Tutor

Extending the Lesson
_____ Applications (Interdisciplinary): CRB page 75
_____ Challenge: SE page 105; CRB page 76 or Internet

ASSESSMENT OPTIONS
_____ Checkpoint Exercises: TE pages 101–102 or Transparencies
_____ Daily Homework Quiz (2.5): TE page 106, CRB page 80, or Transparencies
_____ Standardized Test Practice: SE page 105; TE page 106; STP Workbook; Transparencies
_____ Quiz (2.4–2.5): SE page 106; CRB page 77

Notes _____

_____

_____

TEACHER'S NAME _____ CLASS _____ ROOM _____ DATE _____

Lesson Plan for Block Scheduling

Half-day lesson (See *Pacing the Chapter,* TE pages 64C–64D)   For use with pages 99–107

GOALS 1. **Use a scatter plot to identify the correlation shown by a set of data.**
          2. **Approximate the best-fitting line for a set of data.**

State/Local Objectives _____

_____

_____

✓ **Check the items you wish to use for this lesson.**

CHAPTER PACING GUIDE	
Day	**Lesson**
1	2.1 (all); 2.2 (all)
2	2.3 (all)
3	2.4 (all)
4	**2.5 (all)**; 2.6 (all)
5	2.7 (all); 2.8 (all)
6	Review/Assess Ch. 2

STARTING OPTIONS
____ Homework Check: TE page 95; Answer Transparencies
____ Warm-Up or Daily Homework Quiz: TE pages 100 and 98,
       CRB page 65, or Transparencies

TEACHING OPTIONS
____ Motivating the Lesson: TE page 101
____ Concept Activity: SE page 99; CRB page 66 (Activity Support Master)
____ Lesson Opener (Application): CRB page 67 or Transparencies
____ Graphing Calculator Activity with Keystrokes: CRB page 68
____ Examples 1–3: SE pages 100–102
____ Extra Examples: TE pages 101–102 or Transparencies
____ Technology Activity: SE page 107
____ Closure Question: TE page 102
____ Guided Practice Exercises: SE page 103

APPLY/HOMEWORK
Homework Assignment (See also the assignment for Lesson 2.6.)
____ Block Schedule: 8–24 even, 25–28, 31–43 odd; Quiz 2: 1–9

Reteaching the Lesson
____ Practice Masters: CRB pages 69–71 (Level A, Level B, Level C)
____ Reteaching with Practice: CRB pages 72–73 or Practice Workbook with Examples
____ Personal Student Tutor

Extending the Lesson
____ Applications (Interdisciplinary): CRB page 75
____ Challenge: SE page 105; CRB page 76 or Internet

ASSESSMENT OPTIONS
____ Checkpoint Exercises: TE pages 101–102 or Transparencies
____ Daily Homework Quiz (2.5): TE page 106, CRB page 80, or Transparencies
____ Standardized Test Practice: SE page 105; TE page 106; STP Workbook; Transparencies
____ Quiz (2.4–2.5): SE page 106; CRB page 77

Notes _____

_____

_____

LESSON 2.5

NAME _____ DATE _____

WARM-UP EXERCISES

For use before Lesson 2.5, pages 99–107

1. Find the slope of the line through $(1, 4)$ and $(-5, -2)$.

Find an equation of the line through each pair of points.

2. $(-2, 5)$ and $(1, -1)$

3. $(100, 500)$ and $(150, 1000)$

4. $(0.26, 8.5)$ and $(0.36, 9.8)$

··

DAILY HOMEWORK QUIZ

For use after Lesson 2.4, pages 91–98

1. Write an equation of the line with slope 4 and y-intercept -2.3.

2. Write an equation of the line with slope -1 that passes through $(2, -3)$.

3. Write an equation of the line that passes through $(3, -5)$ and is perpendicular to the line through $(1, 4)$ and $(3, -2)$.

4. Write an equation of the line that passes through $(-2, 5)$ and $(2, -3)$.

The variables x and y vary directly. Write an equation that relates the variables. Then find y when $x = 3$.

5. $x = 4, y = 10$

6. $x = 6, y = \dfrac{1}{2}$

LESSON

2.5

Activity Support Master

For use with page 99

Steps 1 and 2

Distance from projector to screen (cm), x	Length of line segment on screen (cm), y
200	
210	
220	
230	
240	
250	
260	
270	
280	
290	

Steps 3 and 4

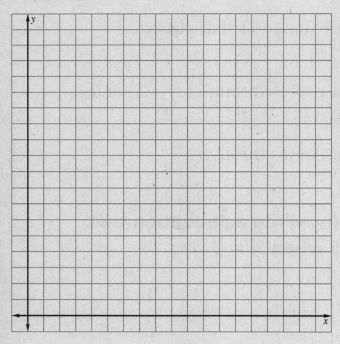

NAME _____ DATE _____

Application Lesson Opener

For use with pages 100–106

1. Are scores on a test and hours studied for the test related? If so, as the number of hours studied increases, would you expect scores to increase or decrease?

Given paired data (x, y), if y tends to increase as x increases then the paired data have a *positive correlation*. If y tends to decrease as x increases then the paired data have a *negative correlation*.

2. Do scores on a test and hours studied have a positive correlation or a negative correlation?

In Questions 3–7, tell whether you expect the paired data to have a *positive correlation* or a *negative correlation*.

3. Hours after a rain shower stops and relative humidity at the location of the shower.

4. Distance traveled in a fixed amount of time and the speed at which a person was traveling.

5. The number of people who attend a movie and the amount of money taken in at the box office.

6. The time it takes to walk a mile and the number of calories burned.

7. The time it takes to walk a mile and the length of the legs of the person who is walking.

LESSON
2.5

NAME _____ DATE _____

Graphing Calculator Activity Keystrokes

For use with page 107

TI-82

STAT 1

Enter *x*-values in L1.

1 **ENTER** 4 **ENTER** 5 **ENTER** 8 **ENTER**

11 **ENTER** 11 **ENTER** 15 **ENTER** 18 **ENTER**

21 **ENTER** 25 **ENTER** 29 **ENTER**

Enter *y*-values in L2.

8 **ENTER** 10 **ENTER** 13 **ENTER** 15 **ENTER**

18 **ENTER** 20 **ENTER** 22 **ENTER** 25 **ENTER**

29 **ENTER** 32 **ENTER** 33 **ENTER**

2nd [STAT PLOT] 1

Choose the following.

On; Type ⸬ ; Xlist L1; Ylist L2; Mark: ▫

WINDOW **ENTER** 0 **ENTER** 30 **ENTER** 5 **ENTER**

0 **ENTER** 35 **ENTER** 5 **ENTER**

GRAPH **STAT** ▶ 5 **2nd** **L1** , **2nd** **L2**

ENTER **Y=** **VARS** 5 ▶ ▶ 7 **GRAPH**

SHARP EL-9600c

STAT [A] **ENTER**

Enter *x*-values in L1.

1 **ENTER** 4 **ENTER** 5 **ENTER** 8 **ENTER**

11 **ENTER** 11 **ENTER** 15 **ENTER** 18 **ENTER**

21 **ENTER** 25 **ENTER** 29 **ENTER**

Enter *y*-values in L2.

8 **ENTER** 10 **ENTER** 13 **ENTER** 15 **ENTER**

18 **ENTER** 20 **ENTER** 22 **ENTER** 25 **ENTER**

29 **ENTER** 32 **ENTER** 33 **ENTER**

2ndF [STAT PLOT] [A] **ENTER**

Choose the following.

on; DATA XY; ListX: L1; ListY: L2.

2ndF [STAT PLOT] [G] 3

ZOOM [A] 9 +/× **CL** **STAT** [D] 0 2 (

2ndF [L1] , **2ndF** [L2] , **VARS** +/× [A]

ENTER 1) **ENTER** **GRAPH**

TI-83

STAT 1

Enter *x*-values in L1.

1 **ENTER** 4 **ENTER** 5 **ENTER** 8 **ENTER**

11 **ENTER** 11 **ENTER** 15 **ENTER** 18 **ENTER**

21 **ENTER** 25 **ENTER** 29 **ENTER**

Enter *y*-values in L2.

8 **ENTER** 10 **ENTER** 13 **ENTER** 15 **ENTER**

18 **ENTER** 20 **ENTER** 22 **ENTER** 25 **ENTER**

29 **ENTER** 32 **ENTER** 33 **ENTER**

2nd [STAT PLOT] 1

Choose the following.

On; Type ⸬ ; Xlist L1; Ylist L2; Mark: ▫

WINDOW 0 **ENTER** 30 **ENTER** 5 **ENTER**

0 **ENTER** 35 **ENTER** 5 **ENTER**

GRAPH **STAT** ▶ 4 **2nd** **L1** , **2nd** **L2**

ENTER **Y=** **VARS** 5 ▶ ▶ 1 **GRAPH**

CASIO CFX-9850GA PLUS

From the main menu, select STAT.

Enter *x*-values in List 1.

1 **EXE** 4 **EXE** 5 **EXE** 8 **EXE** 11 **EXE** 11 **EXE** 15 **EXE**

18 **EXE** 21 **EXE** 25 **EXE** 29 **EXE**

Enter *y*-values in List 2.

8 **EXE** 10 **EXE** 13 **EXE** 15 **EXE** 18 **EXE** 20 **EXE**

22 **EXE** 25 **EXE** 29 **EXE** 32 **EXE** 33 **EXE**

SHIFT **F3** 0 **EXE** 30 **EXE** 5 **EXE** 0 **EXE** 35 **EXE**

5 **EXE** **EXIT**

SHIFT [SetUp] **F2** **EXIT** **F1** **F6**

Choose the following. Graph Type: Scatter; Xlist: List1; Ylist: List2; Frequency: 1; MarkType: ▫

EXIT **F1** **F1** **F6**

LESSON 2.5

NAME _____ DATE _____

Practice A

For use with pages 100–106

Tell whether x and y have a *positive correlation*, a *negative correlation*, or *relatively no correlation*.

1.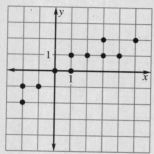

2.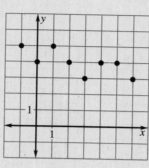

3.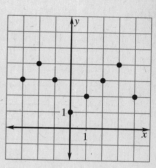

Draw a scatter plot of the data. Then tell whether the data have a *positive correlation*, a *negative correlation*, or *relatively no correlation*.

4.

x	1	2	3	4	5	6	7	8
y	6	6	5	5	6	5	4	3

5.

x	1	2	3	4	5	6	7	8
y	4	5	1	6	6	3	1	6

6.

x	1	2	3	4	5	6	7	8
y	2	2	4	6	8	8	10	10

Approximate the best fitting line for the data.

7.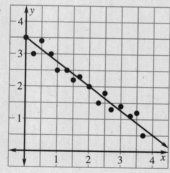

8.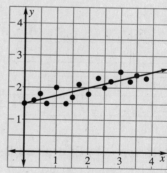

9. *Computers Per Capita* The table shows the number of computers per 1000 people in the U.S. from 1991 through 1995. Draw a scatter plot of the data and describe the correlation shown.

Year	1991	1992	1993	1994	1995
Computers per 1000 people	245.4	266.9	296.6	329.2	364.7

Lesson 2.5

Draw a scatter plot of the data. Then tell whether the data have a *positive correlation,* a *negative correlation,* or *relatively no correlation.*

1.

x	−3	−2.5	−2	−1.75	−1.5	−1	−0.5	0	0.5	0.75	1	1.5
y	0.25	0.5	1	1.5	1.25	2	2.5	2.5	3	3.25	3.5	3.75

2.

x	0	0.5	1	1.25	1.5	2	2.5	3	3.25	3.5	4	4.25
y	2.75	3	2.5	2	1.75	1	1.25	1.5	2.5	3	3.25	3

3.

x	−2	−1	−0.5	0	0.25	1	1.5	2.5	2.75	3.5	4	4.5
y	1	1.25	0.5	0	−1	−1.25	−2	−2.25	−2	−3	−3.25	−3.5

Approximate the best-fitting line for the data.

4.

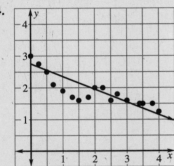

5.

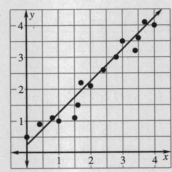

Draw a scatter plot of the data. Then approximate the best-fitting line for the data.

6.

x	−2	−1.5	−1	−0.5	0	0.5	1	1.5	2
y	3	2.5	3	2.4	2.2	2	2.1	1.8	1.5

7.

x	−5	−4	−3	−2	−1	0	1	2	3
y	3	2.5	2.8	3.2	3	4	4.2	4.3	4.5

Broccoli Consumption **In Exercises 8–10, use the following information.**
The table shows the per capita consumption of broccoli, *b* (in pounds), for the years 1980 through 1989.

Year, t	1980	1981	1982	1983	1984	1985	1986	1987	1988	1989
Pounds, b	1.6	1.8	2.2	2.3	2.7	2.9	3.5	3.6	4.2	4.5

8. Draw a scatter plot for the data. Let *t* represent the number of years since 1980.

9. Approximate the best-fitting line for the data.

10. If this pattern were to continue, what would the per capita consumption of broccoli be in 2002?

NAME _____ DATE _____

Practice C

For use with pages 100–106

Approximate the best-fitting line for the data.

1.

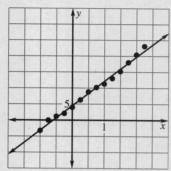

2.

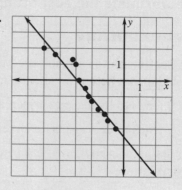

Draw a scatter plot of the data. Then approximate the best-fitting line for the data.

3.

x	−3	−2	−1	0	1	2	3
y	8	7.2	6.4	6	5.5	5	4.8

4.

x	0	1	2	3	4	5	6
y	4	3.5	3.8	4.6	6	7.8	10

5.

x	0	0.5	1	1.5	2	2.5	3
y	0.6	3.2	4.4	5.8	7	8.2	12

African-American Elected Officials **In Exercises 6–8, use the following information.**
The table shows the number of African-American elected officials in U.S. and
state legislatures for the years 1984 to 1993.

Year	1984	1985	1986	1987	1988	1989	1990	1991	1992	1993
Officials	396	407	420	440	436	448	447	484	510	571

6. Draw a scatter plot for the data. Let $t = 4$ represent 1984.

7. Approximate the best-fitting line for the data.

8. If this pattern continues, how many African-American officials will be in
   the U.S. and state legislatures in 2000?

Home Run Champions **In Exercises 9 and 10, use the following information.**
The table shows the number of home runs hit by the American League Home
Run Champion from 1990 to 1996.

Year	1990	1991	1992	1993	1994	1995	1996
Home Runs	51	44	43	46	40	50	52

9. Draw a scatter plot for the data. Let $t = 0$ represent 1990.

10. Approximate the best-fitting line for the data.

NAME _____   DATE _____

Reteaching with Practice

For use with pages 100–106

GOAL **Use a scatter plot to identify the correlation shown by a set of data and approximate the best-fitting line for a set of data**

VOCABULARY

A **scatter plot** is a graph used to determine whether there is a relationship between paired data.

If y tends to increase as x increases, then there is a **positive correlation.**

If y tends to decrease as x increases, then there is a **negative correlation.**

If the points show no linear pattern, then there is **relatively no correlation.**

EXAMPLE 1 *Determining Correlation*

Draw a scatter plot of the data and describe the correlation shown by the scatter plot.

x	-5	-3	-2	-1.5	0	1	2	3	4	5
y	6	5	2	3.5	2	0.5	3	0	-1	1

SOLUTION

The scatter plot shows a negative correlation, which means that as the values of x increase, the values of y tend to decrease.

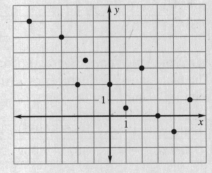

Exercises for Example 1

Draw a scatter plot of the data. Then tell whether the data have a *positive correlation, negative correlation,* or *relatively no correlation.*

1.

x	-2.5	-2	-2	-1	0	0	0	1	2	2
y	-6	-2.5	-4	0	0	-1	-3	3	1	5

2.

x	-3	-3	-2	-1	0	0	2	2	3	5
y	4	0	-1	2	3	-3	-2	4	2	0

NAME _____ DATE _____

Reteaching with Practice

For use with pages 100–106

EXAMPLE 2 *Fitting a Line to Data*

Approximate the best-fitting line for the data in the table.

x	1	1.5	2	3	4	5	6	7	7.5	8
y	7.5	6	6	5	4.5	5	3	3.5	4	3.5

SOLUTION

To begin, draw a scatter plot of the data. Then sketch the line that best fits the points, with as many points above the line as below it.

Now, estimate the coordinates of two points on the line, not necessarily data points. Use these points to find an equation of the line.

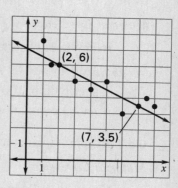

$$m = \frac{y_2 - y_1}{x_2 - x_1} = \frac{3.5 - 6}{7 - 2} = \frac{-2.5}{5} = -\frac{1}{2}$$ Find the slope of the line.

$$y - y_1 = m(x - x_1)$$ Use point-slope form.

$$y - 6 = -\frac{1}{2}(x - 2)$$ Substitute for m, x, and y_1.

$$y - 6 = -\frac{1}{2}x + 1$$ Distributive property

$$y = -\frac{1}{2}x + 7$$ Solve for y.

Exercises for Example 2

Approximate the best-fitting line for the data.

3.

x	−1	−0.5	0.5	1	1.5	2	3	3.5	4	4.2
y	8	8	7	5.5	10	3	3	0.5	0	−2

4. The data in the table shows the age, t (in years), and the corresponding height, h (in inches), for a male from the age of 2 to the age of 19.

Age (t)	2	3	6	8	10	12	14	15	17	18	19
Height (h)	28	33	40	46	52	55	61	64	70	72	72

NAME _____ DATE _____

Quick Catch-Up for Absent Students

For use with pages 99–107

The items checked below were covered in class on (date missed) _____

Activity 2.5: Fitting a Line to a Set of Data (p. 99)

_____ **Goal:** Approximate the best-fitting line for a set of data.

Lesson 2.5: Correlation and Best-Fitting Lines

_____ **Goal 1:** Use a scatter plot to identify the correlation shown by a set of data. (p. 100)

Material Covered:

_____ Example 1: Determining Correlation

Vocabulary:

scatter plot, p. 100   positive correlation, p. 100
negative correlation, p. 100   relatively no correlation, p. 100

_____ **Goal 2:** Approximate the best-fitting line for a set of data. (pp. 101–102)

Material Covered:

_____ Example 2: Fitting a Line to Data

_____ Example 3: Using a Fitted Line

Activity 2.5: Using Linear Regression (p. 107)

_____ **Goal:** Find the best-fitting line for a set of data using the *linear regression* feature of a graphing calculator.

_____ Student Help: Keystroke Help

_____ Other (specify) _____

_____

Homework and Additional Learning Support

_____ Textbook (specify) <u>pp. 103–106</u> _____

_____

_____ *Reteaching with Practice* worksheet (specify exercises) _____

_____ *Personal Student Tutor* for Lesson 2.5

NAME _____ DATE _____

Interdisciplinary Application

For use with pages 100–106

The Universe

SCIENCE   Some scientists believe that the universe is expanding. Edwin Hubble, an astronomer of the early 20th century, collected data on 24 celestial objects that support this theory. When Hubble observed theses 24 celestial objects, he measured how far they were away from Earth, in *megaparsecs*, and how quickly they were receding from Earth, in kilometers per second.

A *parsec* is a unit that astronomers use to measure distance. One parsec is equal to about 19.2 trillion miles, and a *megaparsec* is equal to a thousand parsecs.

In Exercises 1–4, use the following data that was collected by Hubble.

Distance from Earth, x	2.0	2.0	2.0	2.0	1.7	1.4	1.1	1.1	1.0	0.9	0.9
Velocity, y	500	850	800	1090	960	500	450	500	920	−30	650

Distance from Earth, x	0.9	0.9	0.8	0.63	0.5	0.5	0.45	0.275	0.275	0.263
Velocity, y	150	500	300	200	290	270	200	−185	−220	−70

Distance from Earth, x	0.214	0.034	0.032
Velocity, y	−130	290	170

1. Use a graphing calculator to make a scatter plot of the data. Does the scatter plot show a *positive correlation,* a *negative correlation,* or *relatively no correlation?*

2. Use the linear regression feature of a graphing calculator to find an equation of the best-fitting line for the data.

3. Predict the recession velocity of a new object discovered 1.3 megaparsecs from Earth.

4. Suppose you observed an object receding from Earth at the rate of 250 kilometers per second. Predict its distance from Earth.

NAME _____ DATE _____

Challenge: Skills and Applications

For use with pages 100–106

In Exercises 1–3, use the following data on sweater sales at a small store.

Air temperature (°F)	22	28	34	44
Sweaters sold	57	51	38	26

1. The *mean data point* $(\bar{x}, \bar{y})$ for a set of data is the point whose coordinates are the average of all the x-coordinates and the average of all the y-coordinates, respectively.

   a. Calculate the mean data point $(\bar{x}, \bar{y})$ for the data above.

   b. Subtract the coordinates of the mean data point that you found in Part (a) from the coordinates of each data point; that is, subtract the $\bar{x}$ from each x-coordinate and the $\bar{y}$ from each y-coordinate, to get a new table of data points. What is the new mean data point? Make a conjecture based on your answer.

2. a. Calculate the slope m of the best-fitting line for the data above, using the following formula

   $$m = \frac{(x_1 - \bar{x})(y_1 - \bar{y}) + (x_2 - \bar{x})(y_2 - \bar{y}) + \cdots + (x_n - \bar{x})(y_n - \bar{y})}{(x_1 - \bar{x})^2 + (x_2 - \bar{x})^2 + \cdots + (x_n - \bar{x})^2}$$

   Note: You have already calculated the numbers in parentheses in Exercise 1.

   b. The best-fitting line always passes through the point $(\bar{x}, \bar{y})$. Use this fact to write the equation of the best-fitting line for the data above.

3. The *standard deviation* of a set of data is a measure of how much variation there is in the values of a single variable. (Thus, for the data above, there will be a standard deviation for the x-coordinates, denoted σ_x, ("sigma sub x"), and a standard deviation for the y-coordinates, σ_y.)

   a. Use the following formula to find σ_x, for the data above, and the corresponding formula to find σ_y.

   $$\sigma_x = \sqrt{\frac{(x_1 - \bar{x})^2 + (x_2 - \bar{x})^2 + \cdots + (x_n - \bar{x})^2}{n}}$$

   b. Find the standard deviation for the new data you found in part (b) of Exercise 1. Make a conjecture about the standard deviation of such translated data.

LESSON 2.5

Quiz 2

For use after Lessons 2.4–2.5

1. Write an equation of the line that passes through $(0, 4)$ and has a slope of $\frac{3}{4}$. *(Lesson 2.4)*

2. Write an equation of the line that passes through $(-2, -5)$ and has a slope of 4. *(Lesson 2.4)*

3. Write an equation of the line that passes through $(4, -1)$ and has a slope of $-\frac{1}{3}$. *(Lesson 2.4)*

4. Write an equation of the line that passes through $(3, -4)$ and is perpendicular to the line that passes through $(3, 5)$ and $(0, 2)$. *(Lesson 2.4)*

Tell whether *x* and *y* have a *positive correlation*, a *negative correlation*, or *relatively no correlation*. *(Lesson 2.5)*

Answers

1. _____
2. _____
3. _____
4. _____
5. _____
6. _____
7. _____

5.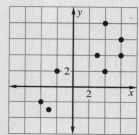

6.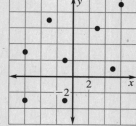

7.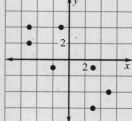

TEACHER'S NAME _____ CLASS _____ ROOM _____ DATE _____

Lesson Plan

1-day lesson (See *Pacing the Chapter,* TE pages 64C–64D)   **For use with pages 108–113**

 GOALS 1. **Graph linear inequalities in two variables.**
2. **Use linear inequalities to solve real-life problems.**

State/Local Objectives _____

_____

✓ Check the items you wish to use for this lesson.

STARTING OPTIONS
____ Homework Check: TE page 103; Answer Transparencies
____ Warm-Up or Daily Homework Quiz: TE pages 108 and 106, CRB page 80, or Transparencies

TEACHING OPTIONS
____ Motivating the Lesson: TE page 109
____ Lesson Opener (Activity): CRB page 81 or Transparencies
____ Examples 1–4: SE pages 108–110
____ Extra Examples: TE pages 109–110 or Transparencies; Internet
____ Closure Question: TE page 110
____ Guided Practice Exercises: SE page 111

APPLY/HOMEWORK
Homework Assignment
____ Basic 14–40 even, 45, 48, 49, 52, 57–73 odd
____ Average 14–44 even, 45–49, 52, 57–73 odd
____ Advanced 14–44 even, 45–55, 57–73 odd

Reteaching the Lesson
____ Practice Masters: CRB pages 82–84 (Level A, Level B, Level C)
____ Reteaching with Practice: CRB pages 85–86 or Practice Workbook with Examples
____ Personal Student Tutor

Extending the Lesson
____ Applications (Real-Life): CRB page 88
____ Challenge: SE page 113; CRB page 89 or Internet

ASSESSMENT OPTIONS
____ Checkpoint Exercises: TE pages 109–110 or Transparencies
____ Daily Homework Quiz (2.6): TE page 113, CRB page 92, or Transparencies
____ Standardized Test Practice: SE page 113; TE page 113; STP Workbook; Transparencies

Notes _____

_____

_____

Algebra 2
Chapter 2 Resource Book

TEACHER'S NAME _____ CLASS _____ ROOM _____ DATE _____

Lesson Plan for Block Scheduling

Half-day lesson (See *Pacing the Chapter*, TE pages 64C–64D)   For use with pages 108–113

GOALS   1. Graph linear inequalities in two variables.
2. Use linear inequalities to solve real-life problems.

State/Local Objectives _____

_____

_____

✓ **Check the items you wish to use for this lesson.**

STARTING OPTIONS
____ Homework Check: TE page 103; Answer Transparencies
____ Warm-Up or Daily Homework Quiz: TE pages 108 and 106,
       CRB page 80, or Transparencies

TEACHING OPTIONS
____ Motivating the Lesson: TE page 109
____ Lesson Opener (Activity): CRB page 81 or Transparencies
____ Examples 1–4: SE pages 108–110
____ Extra Examples: TE pages 109–110 or Transparencies; Internet
____ Closure Question: TE page 110
____ Guided Practice Exercises: SE page 111

APPLY/HOMEWORK
Homework Assignment (See also the assignment for Lesson 2.5.)
____ Block Schedule: 14–44 even, 45–49, 52, 57–73 odd

Reteaching the Lesson
____ Practice Masters: CRB pages 82–84 (Level A, Level B, Level C)
____ Reteaching with Practice: CRB pages 85–86 or Practice Workbook with Examples
____ Personal Student Tutor

Extending the Lesson
____ Applications (Real Life): CRB page 88
____ Challenge: SE page 113; CRB page 89 or Internet

ASSESSMENT OPTIONS
____ Checkpoint Exercises: TE pages 109–110 or Transparencies
____ Daily Homework Quiz (2.6): TE page 113, CRB page 92, or Transparencies
____ Standardized Test Practice: SE page 113; TE page 113; STP Workbook; Transparencies

Notes _____

_____

_____

CHAPTER PACING GUIDE	
Day	Lesson
1	2.1 (all); 2.2 (all)
2	2.3 (all)
3	2.4 (all)
4	2.5 (all); **2.6 (all)**
5	2.7 (all); 2.8 (all)
6	Review/Assess Ch. 2

Lesson 2.6

NAME _____ DATE _____

WARM-UP EXERCISES

For use before Lesson 2.6, pages 108–113

Graph each line on the same coordinate grid.

1. $y = 2$

2. $x = -1$

3. $-x + y = 1$

4. $y = 3x + 1$

DAILY HOMEWORK QUIZ

For use after Lesson 2.5, pages 99–107

1. Draw a scatter plot of the data. Tell whether x and y have a *positive correlation,* a *negative correlation,* or *relatively no correlation.*

   $(-3, 1), (-3, 2), (-2, 0), (-1, 2), (0, 0),$

   $(1, -2), (2, -1), (2, -2), (3, -1), (3, -3)$

2. Draw a scatter plot of the data. Approximate the best-fitting line.

   $(2, 10), (2, 12), (4, 11), (6, 7), (6, 9),$

   $(8, 6), (8, 8), (10, 6), (12, 3), (12, 6)$

3. Use the results from Exercise 2 to predict the value of y when $x = 14.$

Algebra 2
Chapter 2 Resource Book

NAME _____ DATE _____

Activity Lesson Opener

For use with pages 108–113

SET UP: Work in a group.

YOU WILL NEED: • colored pencils • ruler

Use the inequality that is assigned to you.

Group 1: $y \geq 2x - 3$ Group 2: $y \leq 2x - 3$

Group 3: $y \leq -2x + 3$ Group 4: $y \geq -2x - 3$

Group 5: $y < \frac{1}{2}x + 2$ Group 6: $y > \frac{1}{2}x - 2$

Group 7: $y > -\frac{1}{2}x + 2$ Group 8: $y > -\frac{1}{2}x - 2$

Group 9: $y > 2x - 4$ Group 10: $y < -2x + 4$

Your teacher will call out ordered pairs randomly generated by flipping a coin and rolling a number cube twice. Heads indicate the number on the number cube is positive and tails indicate it is negative. A 6 on a number cube will be a zero.

1. Decide if the ordered pair is a solution to the inequality. If it is, plot it on the grid with a red pencil. If it is not, plot it with a blue pencil.

After ten points are called out, the group with the most red points wins.

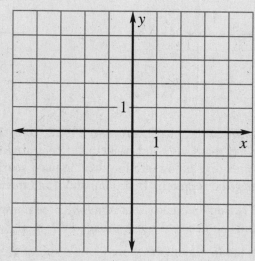

2. The boundary of the solution region is a line. Is the boundary line included in the region? If it is, graph it with a red pencil. If not, graph it with a blue pencil. Finish graphing the inequality by shading the entire solution with a red pencil.

NAME _____ DATE _____

Practice A

For use with pages 108–113

Check whether the given ordered pairs are solutions of the inequality.

1. $x + y < 5$; $(1, 2), (7, -2)$

2. $x > 3$; $(0, 4), (5, 1)$

3. $y \leq -1$; $(-1, 3), (2, -1)$

4. $y - x \geq -1$; $(5, 6), (-3, -1)$

5. $x < 2y + 5$; $(4, 0), (-4, -5)$

6. $y \geq x - 7$; $(2, 4), (8, -3)$

Graph the inequality in a coordinate plane.

7. $x > -3$

8. $x < 1$

9. $x \geq 5$

10. $x \leq -7$

11. $y > 1$

12. $y < -6$

13. $y \leq -2$

14. $y \geq 4$

Match the inequality with its graph.

15. $3x + y > 1$

16. $2x - y \leq -3$

17. $-4x + y < -1$

18. $-2x + y \leq 0$

19. $5x > 2$

20. $3y < 6$

A.
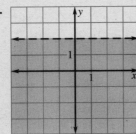

B.
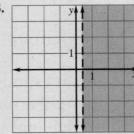

C.
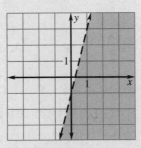

D.
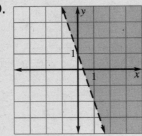

E.
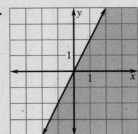

F.
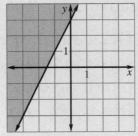

Basketball Stats **In Exercises 21–23, use the following information.**

In order for this year's star basketball player to break the school record for most points (excluding free throws), he must score at least 34 points. The points may be scored by two-point shots and three-point shots.

21. Write an inequality that represents the number of two- and three-point shots he needs to break the record.

22. In the first game he scored 13 two-point shots and 2 three-point shots. Did he break the record?

23. Give two possible combinations of two- and three-point shots that will give him the record.

NAME _____ DATE _____

Practice B

For use with pages 108–113

Check whether the given ordered pairs are solutions of the inequality.

1. $2x - 3y \le 2$; $(0, -1), (3, 2)$

2. $x + 2y > 4$; $(2, 1), (-3, 6)$

3. $5x + y \ge -3$; $(-3, 6), (2, -5)$

4. $3x - 10y < -8$; $(6, 3), (-4, -2)$

5. $4y - 2x < 5$; $(2, 0), (-3, 1)$

6. $2y + x \ge 3$; $(-1, -2), (1, 1)$

Graph the inequality in a coordinate plane.

7. $x \ge 1$

8. $x < -\dfrac{1}{2}$

9. $2x > 6$

10. $y < 4$

11. $y \ge -5$

12. $\dfrac{1}{3}y \ge -2$

13. $y < 2x - 1$

14. $y \ge \dfrac{1}{2}x + 5$

15. $4x + y \le -2$

16. $x + 2y > 4$

17. $-5x + 5y > 1$

18. $3x - y \le 7$

19. $2x - 4y > 8$

20. $6x - 3y \ge -1$

21. $12x + 4y < 8$

Defrosting Meat **In Exercises 22–24, use the following information.**
According to one cookbook, you should always defrost meat in the original
wrappings on a refrigerator shelf. You should allow 5 hours for each pound, less
for thinner cuts.

22. Write and graph an inequality that represents the time t (in hours) and the
number of pounds p of meat being defrosted. Use t on the vertical axis
and p on the horizontal axis.

23. What are the coordinates of a 2-pound roast that has been defrosting for
12 hours?

24. Is it possible that the roast in Exercise 23 is completely defrosted? Explain
your answer.

Fundraiser **In Exercises 25–27, use the following information.**
An environmentalist group is planning a fundraiser. The group wants to purchase caps and T-shirts with their logo on them and sell them at a profit. They
can buy caps for $3 each and T-shirts for $5 each. They have $800 to spend.

25. Write and graph an inequality that represents the numbers of caps x and
T-shirts y that the group can buy.

26. Suppose the group purchased 50 caps and 150 T-shirts. What point on
the coordinate plane represents this purchase?

27. Is the point in Exercise 26 a solution of the inequality?

NAME _____ DATE _____

Practice C
For use with pages 108–113

Graph the inequality in a coordinate plane.

1. $x - 3 < 5$

2. $y + 2 > -3$

3. $-3x + 2y \geq 0$

4. $-4x + 7y > 0$

5. $2x + 3y \leq 6$

6. $4x - 3y > 12$

7. $3x - 2y \geq 9$

8. $-5x + 3y < 10$

9. $7x + 4y \leq 8$

10. $6x - 5y > 10$

11. $4x + 3y \geq 2$

12. $8x - 9y \leq 3$

13. $2x + 3y < 5$

14. $4x - 3y > 1$

15. $3x + 5y \leq 8$

Test Scores   **In Exercises 16–18, use the following information.**

A history exam included multiple choice questions that were worth 4 points each and true/false questions that were worth 2 points each. The highest score earned by a person in your class was 92.

16. Write an inequality that represents the number of multiple choice questions and true/false questions that could have been answered correctly by any member of your class.

17. Graph the inequality.

18. Is it possible that someone answered 20 multiple choice questions and 7 true/false questions correctly?

Babysitting Wages   **In Exercises 19 and 20, use the following information.**

You earn $3 per hour when you babysit the Thompson children. You earn $3.50 per hour when you babysit the Stewart children. You would like to buy a $47.50 ticket for a concert that is coming to town in 5 weeks.

19. Write and graph an inequality that represents the number of hours you need to babysit for the Thompson's and Stewart's to earn enough money to buy your concert ticket.

20. Give three possible combinations of babysitting hours that satisfy the inequality.

Visual Thinking   Write the inequality represented by the graph.

21.
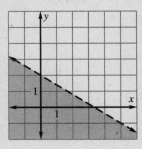

22.
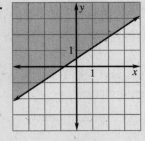

Algebra 2
Chapter 2 Resource Book

NAME _____ DATE _____

Reteaching with Practice

For use with pages 108–113

GOAL **Graph linear inequalities in two variables and use linear inequalities to solve real-life problems**

VOCABULARY

A **linear inequality** in two variables can be written in one of the following forms: $Ax + By < C$, $Ax + By \leq C$, $Ax + By > C$, $Ax + By \geq C$.

An ordered pair (x, y) is a **solution** of a linear inequality if the inequality is true when the values for x and y are substituted into the inequality.

A **graph** of a linear inequality in two variables is the graph of all solutions of the inequality.

The boundary line of the inequality divides the coordinate plane into two **half-planes;** a shaded region containing the points that are solutions of the inequality, and an unshaded region which contains the points that are not.

EXAMPLE 1 *Checking Solutions of Inequalities*

Check whether the given ordered pair is a solution of $-x + 2y < 6$.

a. $(0, -6)$ **b.** $(2, 4)$ **c.** $(-3, 2)$

SOLUTION

Ordered Pair	Substitute	Conclusion
a. $(0, -6)$	$-(0) + 2(-6) = -12 < 6$	$(0, -6)$ is a solution.
b. $(2, 4)$	$-(2) + 2(4) = 6 \not< 6$	$(2, 4)$ is not a solution.
c. $(-3, 2)$	$-(-3) + 2(2) = 7 \not< 6$	$(-3, 2)$ is not a solution.

Exercises for Example 1

Check whether the given ordered pairs are solutions of the inequality.

1. $x \geq -1$; $(-1, -2), (5, 2)$ **2.** $y \leq x + 1$; $(4, 5), (-2, 1)$

3. $y > 5$; $(2, 6), (0, 2)$ **4.** $4x < -9$; $(3, -2), (-2, -5)$

EXAMPLE 2 *Graphing Linear Inequalities in One Variable*

Graph (a) $2y > 6$ and (b) $x \geq -5$ in the coordinate plane.

SOLUTION

a. Graph the boundary line $y = 3$. Use a dashed line because $y > 3$.

b. Graph the boundary line $x = -5$. Use a solid line because $x \geq -5$.

NAME _____ DATE _____

Reteaching with Practice

For use with pages 108–113

Test the point $(0, 0)$. Because $(0, 0)$ is *not* a solution of the inequality, shade the half-plane above the line.

Test the point $(0, 0)$. Because $(0, 0)$ *is* a solution of the inequality, shade the half-plane to the right of the line.

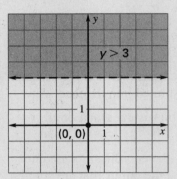

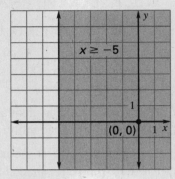

Exercises for Example 2

Graph the inequality in the coordinate plane.

5. $2x \geq -4$

6. $-y < -2$

7. $3x \leq 9$

EXAMPLE 3 *Graphing Linear Inequalities in Two Variables*

Graph $y > -2x - 5$.

SOLUTION

Graph the boundary line $y = -2x - 5$. Use a dashed line because $y > -2x - 5$. Test the point $(0, 0)$. Because $(0, 0)$ *is* a solution of the inequality, shade the half-plane above the line.

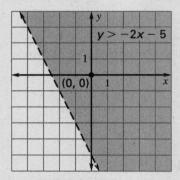

Exercises for Example 3

Graph the inequality.

8. $y \geq -x + 2$

9. $y < -\frac{1}{2}x + 4$

10. $-y \leq x + 3$

NAME _____ DATE _____

Quick Catch-Up for Absent Students

For use with pages 108–113

The items checked below were covered in class on (date missed) _____

Lesson 2.6: Linear Inequalities in Two Variables

_____ **Goal 1:** Graph linear inequalities in two variables. (pp. 108–109)

Material Covered:

_____ Example 1: Checking Solutions of Inequalities

_____ Activity: Investigating the Graph of an Inequality

_____ Student Help: Look Back

_____ Example 2: Graphing Linear Inequalities in One Variable

_____ Student Help: Study Tip

_____ Example 3: Graphing Linear Inequalities in Two Variables

Vocabulary:

      linear inequality in two variables, p. 108

      solution of a linear inequality in two variables, p. 108

      graph of a linear inequality in two variables, p. 108

      half-plane, p. 108

_____ **Goal 2:** Use linear inequalities to solve real-life problems. (p. 110)

Material Covered:

_____ Example 4: Writing and Using a Linear Inequality

_____ Other (specify) _____

_____

Homework and Additional Learning Support

_____ Textbook (specify) <u>pp. 111–113</u> _____

_____

_____ Internet: Extra Examples at www.mcdougallittell.com

_____ *Reteaching with Practice* worksheet (specify exercises) _____

_____ *Personal Student Tutor* for Lesson 2.6

NAME _____ DATE _____

Real–Life Application:
When Will I Ever Use This?

For use with pages 108–113

Playing Computer Games

Some of the more popular computer games are called simulations. They generally take a real-life situation and reproduce it in an exciting and entertaining way.

Suppose you are playing a game that simulates the workings of an ancient tribe of people. These people have to support their village by obtaining food from the wild. Since they are rather primitive, they only know of two ways of getting food: hunting and gathering. You are going to use your knowledge of linear inequalities to direct your people in the most efficient way to gather food.

Suppose your tribe of people has ten people designated to gather food, with the following conditions.

- Each one of them can gather food for up to twelve hours a day.

- Each hour hunting will yield six units of food, while each hour gathering will yield four.

- Tribal customs require that for each hour hunting, at least two hours must be spent gathering.

Answer the following questions to optimize the amount of food that your people can gather.

1. How many hours total can the tribe devote to obtaining food each day?

2. Let *x* represent hours spent hunting and *y* represent hours spent gathering. Write an inequality to represent how these quantities relate to the total number of hours spent obtaining food.

3. Write an inequality using *x* and *y* to represent the restriction due to tribal customs.

4. Knowing that each hour hunting will yield six units of food, while each hour gathering will yield four, write an expression using *x* and *y* to represent how much food is collected.

5. Graph the inequalities from Exercises 2 and 3. Also, graph the inequalities $x \geq 0$ and $y \geq 0$ (because these variables can't be less than zero—you cannot work negative hours!). Shade in the areas that *all* the inequalities share. Your shaded area should look like a polygon.

6. Pick four ordered pairs inside the region you have just shaded, and evaluate them in the expression you found in Exercise 4. Which pair yields the most food?

7. Compare your answers with others in the class, and see who has the best scheme for gathering food.

NAME _____ DATE _____

Challenge: Skills and Applications

For use with pages 108–113

1. In the graph at the right, the solid line is the graph of $y = \frac{1}{2}x + 3$.

   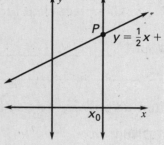

   a. What is the equation of the vertical line shown? In terms of x_0, what is the y-coordinate of the point where it intersects the graph?

   b. Write an inequality in one variable whose solutions are precisely the y-coordinates of the points on the vertical line and *above* the graph.

   c. Based on your answers to parts (a) and (b), describe a method of graphing an inequality in two variables that does not require testing any points.

In Exercises 2 and 3, graph each inequality.

2. $2(x - y) < 3x + y + 2$

3. $\dfrac{2x + 4}{y} \le 5$ (*Hint*: Graph the equality, then consider the inequality cases $y > 0$ and $y < 0$ separately.)

4. **a.** Describe in words the family of lines whose questions are of the form $y - 2 = m(x - 5)$, for all possible values of m.

   b. For what values of m will the lines have a y-intercept that is greater than or equal to 0?

   c. For what values of m will the lines have an x-intercept that is greater than or equal to 0? For what values of m will both intercepts be positive?

5. Suppose a line is defined by the parametric equations

   $x = a + 2t$

   $y = b + 3t.$

   a. By solving for t in the first equation and substituting the expression you get into the second equation, write the equation of the line in point-slope form, in terms of the constants a and b.

   b. Suppose you know that $a \le 4$ and $b \ge 5$. Write a single inequality in x and y that describes the half-plane consisting of the point lines satisfying the parametric equations in part (a).

TEACHER'S NAME _____ CLASS _____ ROOM _____ DATE _____

Lesson Plan

1-day lesson (See *Pacing the Chapter,* TE pages 64C–64D)   **For use with pages 114–121**

 GOALS
1. **Represent piecewise functions.**
2. **Use piecewise functions to model real-life quantities.**

State/Local Objectives _____

_____

✓ Check the items you wish to use for this lesson.

STARTING OPTIONS
____ Homework Check: TE page 111; Answer Transparencies
____ Warm-Up or Daily Homework Quiz: TE pages 114 and 113, CRB page 92, or Transparencies

TEACHING OPTIONS
____ Lesson Opener (Application): CRB page 93 or Transparencies
____ Graphing Calculator Activity with Keystrokes: CRB pages 94–95
____ Examples 1–6: SE pages 114–116
____ Extra Examples: TE pages 115–116 or Transparencies
____ Technology Activity: SE page 121
____ Closure Question: TE page 116
____ Guided Practice Exercises: SE page 117

APPLY/HOMEWORK
Homework Assignment
____ Basic 14–26 even, 27–41 odd, 60, 61, 63–71 odd
____ Average 14–26 even, 27–43 odd, 50–52, 60, 61, 63–71 odd
____ Advanced 14–26 even, 27–47 odd, 50–55, 60–62, 63–71 odd

Reteaching the Lesson
____ Practice Masters: CRB pages 96–98 (Level A, Level B, Level C)
____ Reteaching with Practice: CRB pages 99–100 or Practice Workbook with Examples
____ Personal Student Tutor

Extending the Lesson
____ Applications (Interdisciplinary): CRB page 102
____ Challenge: SE page 120; CRB page 103 or Internet

ASSESSMENT OPTIONS
____ Checkpoint Exercises: TE pages 115–116 or Transparencies
____ Daily Homework Quiz (2.7): TE page 120, CRB page 106, or Transparencies
____ Standardized Test Practice: SE page 120; TE page 120; STP Workbook; Transparencies

Notes _____

_____

_____

TEACHER'S NAME _____ CLASS _____ ROOM _____ DATE _____

Lesson Plan for Block Scheduling

Half-day lesson (See *Pacing the Chapter*, TE pages 64C–64D)   For use with pages 114–121

GOALS 1. **Represent piecewise functions.**
2. **Use piecewise functions to model real-life quantities.**

State/Local Objectives _____

_____

_____

✓ **Check the items you wish to use for this lesson.**

STARTING OPTIONS

_____ Homework Check: TE page 111; Answer Transparencies
_____ Warm-Up or Daily Homework Quiz: TE pages 114 and 113,
        CRB page 92, or Transparencies

TEACHING OPTIONS

_____ Lesson Opener (Application): CRB page 93 or Transparencies
_____ Graphing Calculator Activity with Keystrokes: CRB pages 94–95
_____ Examples 1–6: SE pages 114–116
_____ Extra Examples: TE pages 115–116 or Transparencies
_____ Technology Activity: SE page 121
_____ Closure Question: TE page 116
_____ Guided Practice Exercises: SE page 117

APPLY/HOMEWORK

Homework Assignment (See also the assignment for Lesson 2.8.)
_____ Block Schedule: 14–26 even, 27–43 odd, 50–52, 60, 61, 63–71 odd

Reteaching the Lesson
_____ Practice Masters: CRB pages 96–98 (Level A, Level B, Level C)
_____ Reteaching with Practice: CRB pages 99–100 or Practice Workbook with Examples
_____ Personal Student Tutor

Extending the Lesson
_____ Applications (Interdisciplinary): CRB page 102
_____ Challenge: SE page 120; CRB page 103 or Internet

ASSESSMENT OPTIONS

_____ Checkpoint Exercises: TE pages 115–116 or Transparencies
_____ Daily Homework Quiz (2.7): TE page 120, CRB page 106, or Transparencies
_____ Standardized Test Practice: SE page 120; TE page 120; STP Workbook; Transparencies

Notes _____

_____

_____

CHAPTER PACING GUIDE

Day	Lesson
1	2.1 (all); 2.2 (all)
2	2.3 (all)
3	2.4 (all)
4	2.5 (all); 2.6 (all)
5	**2.7 (all)**; 2.8 (all)
6	Review/Assess Ch. 2

Lesson 2.7

WARM-UP EXERCISES

For use before Lesson 2.7, pages 114–121

1. Evaluate $f(x) = 3x - 2$ when $x = -2$.

2. Evaluate $h(x) = \dfrac{3}{2}x + \dfrac{5}{2}$ when $x = -5$.

3. Graph the lines $-x + y = -2$ and $y = \dfrac{1}{3}x - 2$ on the same coordinate grid.

DAILY HOMEWORK QUIZ

For use after Lesson 2.6, pages 108–113

Graph the inequality in a coordinate plane.

1. $2y \le 6$

2. $x + 2y > 2$

3. $1.5x - 3y \le -9$

NAME _____ DATE _____

Application Lesson Opener

For use with pages 114–120

**Center High School held a four-hour fundraising pledge
drive. The students organizing the drive counted the total
money raised at the end of each hour. The results are shown
in the graph.**

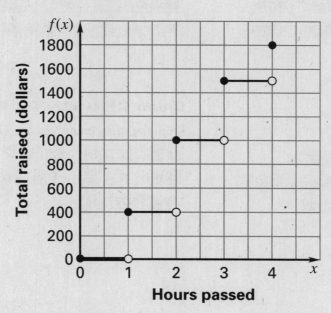

1. How much money had the students raised after 2 hours?

2. How much money did they raise in all?

3. If x is the number of hours of the drive that have passed, then
   the function $f(x)$ shown by the graph gives the number of dollars
   raised. What is $f(3.5)$?

4. On what interval does $f(x) = 400$?

5. What is the domain of the function?

6. What is the range of the function?

Lesson 2.7

Graphing Calculator Activity Keystrokes

For use with page 118

Keystrokes for Exercise 41

TI-82

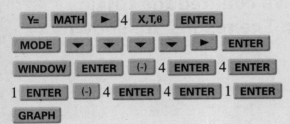

| Y= | MATH | ► | 4 | X,T,θ | ENTER |

MODE ▼ ▼ ▼ ▼ ► ENTER

WINDOW ENTER (-) 4 ENTER 4 ENTER

1 ENTER (-) 4 ENTER 4 ENTER 1 ENTER

GRAPH

TI-83

Y= MATH ► 5 X,T,θ,n) ENTER

MODE ▼ ▼ ▼ ▼ ► ENTER

WINDOW (-) 4 ENTER 4 ENTER 1 ENTER

(-) 4 ENTER 4 ENTER 1 ENTER GRAPH

SHARP EL-9600c

Insert function

Y= MATH [B] 5 X/θ/T/n ENTER

WINDOW (-) 4 ENTER 4 ENTER 1 ENTER

(-) 4 ENTER 4 ENTER 1 ENTER

GRAPH

CASIO CFX-9850Ga PLUS

From the main menu, choose GRAPH.

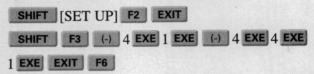

SHIFT [SET UP] F2 EXIT

SHIFT F3 (-) 4 EXE 1 EXE (-) 4 EXE 4 EXE

1 EXE EXIT F6

Graphing Calculator Activity Keystrokes

TI-82

[MODE]

Choose the following.

Normal; Float; Degree; Function; Dot; Sequential; Full Screen.

[Y=] [( ] 3 [X,T,θ] [−] 1 [ )] [( ] [X,T,θ] [2nd] [TEST] 5 2 [ )] [+] [( ] 7 [ )] [( ] 2 [2nd] [TEST] 6 [X,T,θ] [2nd] [TEST] [►] 1 [X,T,θ] [2nd] [TEST] 6 5 [ )] [+] [( ] 2 [X,T,θ] [−] 3 [ )] [( ] [X,T,θ] [2nd] [TEST] 3 5 [ )]

Adjust window and graph

[WINDOW] [ENTER] [(-)] 5 [ENTER] 15 [ENTER] 1 [ENTER] [(-)] 1 [ENTER] 20 [ENTER] 1 [ENTER] [GRAPH]

Press [TRACE] and use the left and right arrow keys to move the trace cursor. See that $f(4) \approx 7$ and $f(10) \approx 17$.

TI-83

[MODE]

Choose the following.

Normal; Float; Degree; Function; Dot; Sequential; Real; Full.

[Y=] [( ] 3 [X,T,θ,n] [−] 1 [ )] [( ] [X,T,θ,n] [2nd] [TEST] [5] 2 [ )] [+] [( ] 7 [ )] [( ] 2 [2nd] [TEST] [6] [X,T,θ,n] [2nd] [TEST] [►] [1] [X,T,θ,n] [2nd] [TEST] [6] 5 [ )] [+] [( ] 2 [X,T,θ,n] [−] 3 [ )] [( ] [X,T,θ,n] [2nd] [TEST] [3] 5 [ )]

Adjust window and graph

[WINDOW] [(-)] 5 [ENTER] 15 [ENTER] 1 [ENTER] [(-)] 1 [ENTER] 20 [ENTER] 1 [ENTER] 1 [ENTER] [GRAPH]

Press [TRACE] and use the left and right arrow keys to move the trace cursor. See that $f(4) \approx 7$ and $f(10) \approx 17$.

SHARP EL-9600c

Select dot mode.

[2ndF] [FORMAT] [E] 2 [Y=] [( ] 3 [X/θ/T/n] [−] 1 [ )] [( ] [X/θ/T/n] [MATH] [F] 5 2 [ )] [ENTER] [( ] 7 [ )] [( ] 2 [MATH] [F] 6 [X/θ/T/n] [ )] [MATH] [G] 1 [( ] 7 [ )] [( ] [X/θ/T/n] [MATH] [F] 6 5 [ )] [ENTER] [( ] 2 [X/θ/T/n] [−] 3 [ )] [( ] [X/θ/T/n] [MATH] [F] 3 5 [ )] [ENTER]

[WINDOW] [(-)] 5 [ENTER] 15 [ENTER] 1 [ENTER] [(-)] 1 [ENTER] 20 [ENTER] 1 [ENTER] [GRAPH]

Press [TRACE] and use the left and right arrow keys to move the trace cursor. Use the up and down arrow keys to select a different branch of the function f. See that $f(4) \approx 7$ and that $f(10) \approx 17$.

[2ndF] [CALC] [1] 4 [ENTER] [2nd] [CALC] [1] 10 [ENTER]

CASIO CFX-9850GA PLUS

From the main menu, select GRAPH.

Select dot mode.

[SHIFT] [SET UP] [F2] [EXIT] 3 [X,θ,T] [−] 1 [,] [SHIFT] [[ ] [(-)] 5 [,] 2 [SHIFT] [ ]] [EXE] 7 [,] [SHIFT] [[ ] 2 [,] 5 [SHIFT] [ ]] [EXE] 2 [X,θ,T] [−] 3 [,] [SHIFT] [[ ] 5 [,] 15 [SHIFT] [ ]] [EXE] [SHIFT] [F3] [(-)] 5 [EXE] 15 [EXE] 1 [EXE] [(-)] 1 [EXE] 20 [EXE] 1 [ENTER] [EXIT] [F6]

Press [F1] and use the left and right arrow keys to move the trace cursor. Use the up and down arrow keys to select a different branch of the function f. See that $f(4) \approx 7$ and that $f(10) \approx 17$.

Lesson 2.7

NAME _____ DATE _____

Practice A
For use with pages 114–120

Evaluate the function for the given value of *x*.

$$f(x) = \begin{cases} 3, & \text{if } x \leq 0 \\ 2, & \text{if } x > 0 \end{cases}$$

$$g(x) = \begin{cases} x + 5, & \text{if } x \leq 3 \\ 2x - 1, & \text{if } x > 3 \end{cases}$$

$$h(x) = \begin{cases} \frac{1}{2}x - 4, & \text{if } x \leq -2 \\ 3 - 2x, & \text{if } x > -2 \end{cases}$$

1. $f(2)$

2. $f(-4)$

3. $f(0)$

4. $f\left(\frac{1}{2}\right)$

5. $g(7)$

6. $g(0)$

7. $g(-1)$

8. $g(3)$

9. $h(-4)$

10. $h(-2)$

11. $h(-1)$

12. $h(6)$

Match the piecewise function with its graph.

13. $f(x) = \begin{cases} x - 4, & \text{if } x \leq 1 \\ 3x, & \text{if } x > 1 \end{cases}$

14. $f(x) = \begin{cases} x + 4, & \text{if } x \leq 0 \\ 2x + 4, & \text{if } x > 0 \end{cases}$

15. $f(x) = \begin{cases} 3x - 2, & \text{if } x \leq 1 \\ x + 2, & \text{if } x > 1 \end{cases}$

16. $f(x) = \begin{cases} 2x + 3, & \text{if } x \geq 0 \\ x + 4, & \text{if } x < 0 \end{cases}$

17. $f(x) = \begin{cases} 3x - 1, & \text{if } x \geq -1 \\ -5, & \text{if } x < -1 \end{cases}$

18. $f(x) = \begin{cases} -3x - 1, & \text{if } x \leq 1 \\ -5, & \text{if } x > 1 \end{cases}$

A.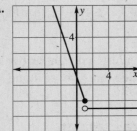

B.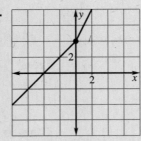

C.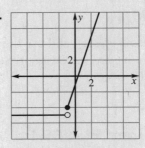

D.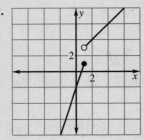

E.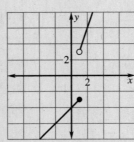

F.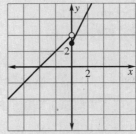

19. *Amusement Park Rates* The admission rates at an amusement park are as follows.

> Children 5 years old and under: free
> Children over 5 years and up to (and including) 12 years: $5.00
> Children over 12 years and up to (and including) 18 years: $12.00
> Adults: $18.00

Write a piecewise function that gives the admission price for a given age. Graph the function.

NAME _____ DATE _____

Practice B

For use with pages 114–120

Evaluate the function for the given value of *x*.

$$f(x) = \begin{cases} 3x - 7, & \text{if } x \le 2 \\ 6 - 2x, & \text{if } x > 2 \end{cases} \qquad g(x) = \begin{cases} 3x + 5, & \text{if } x < 5 \\ -x + 3, & \text{if } x \ge 5 \end{cases} \qquad h(x) = \begin{cases} \frac{2}{3}x + 1, & \text{if } x > -3 \\ 2x - 3, & \text{if } x \le -3 \end{cases}$$

1. $f(0)$ **2.** $f(2)$ **3.** $f(4)$ **4.** $f(-3)$

5. $g(5)$ **6.** $g(-4)$ **7.** $g(3)$ **8.** $g(10)$

9. $h(-9)$ **10.** $h(-3)$ **11.** $h(6)$ **12.** $h(1)$

Graph the function.

13. $f(x) = \begin{cases} 3, & \text{if } x \le 4 \\ -1, & \text{if } x > 4 \end{cases}$ **14.** $f(x) = \begin{cases} x + 3, & \text{if } x \le 0 \\ 2x, & \text{if } x > 0 \end{cases}$ **15.** $f(x) = \begin{cases} x - 4, & \text{if } x < 2 \\ 3 - x, & \text{if } x \ge 2 \end{cases}$

16. $f(x) = \begin{cases} 2x + 3, & \text{if } x \ge -1 \\ -3x + 1, & \text{if } x < -1 \end{cases}$ **17.** $f(x) = \begin{cases} -x, & \text{if } x > 5 \\ \frac{2}{5}x, & \text{if } x \le 5 \end{cases}$ **18.** $f(x) = \begin{cases} \frac{1}{2} - x, & \text{if } x > 0 \\ 2x + 3, & \text{if } x \le 0 \end{cases}$

19. $f(x) = \begin{cases} x + 1, & \text{if } x < 0 \\ -x + 1, & \text{if } 0 \le x \le 2 \\ x - 1, & \text{if } x > 2 \end{cases}$ **20.** $f(x) = \begin{cases} 2x, & \text{if } x \ge -1 \\ 3x, & \text{if } -2 < x < -1 \\ -x, & \text{if } x \le -2 \end{cases}$ **21.** $f(x) = \begin{cases} 2, & \text{if } x \le -3 \\ -1, & \text{if } -3 < x < 3 \\ 3, & \text{if } x \ge 3 \end{cases}$

Write equations for the piecewise function whose graph is shown.

22.

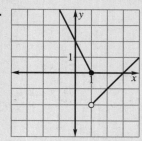

23.

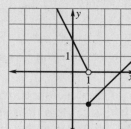

24.

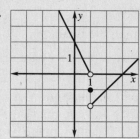

Tour Bus   **In Exercises 25 and 26, use the following information.**

A company provides bus tours of historical cities. The given function describes
the rate for small groups and the discounted rate for larger groups, where *x* is the
number of people in your group.

$$C = \begin{cases} 8.95x, & \text{if } 0 < x \le 10 \\ 7.50x, & \text{if } x > 10 \end{cases}$$

25. Graph the function.

26. Identify the domain and range of the function.

27. ***Commission Rate***   You are employed by a company in which commission
rates are based on how much you sell. If you sell up to $100,000 of mer-
chandise in a month, you earn 5% of sales as a commission. If you sell
over $100,000, you earn 8% commission on your sales. Write a piecewise
function that gives the amount you earn in commission in a given month
for *x* dollars in sales.

NAME _____ DATE _____

Practice C

For use with pages 114–120

Evaluate the function for the given value of x.

$$f(x) = \begin{cases} 3x + 5, & \text{if } x < \frac{1}{2} \\ 2x + 1, & \text{if } x \geq \frac{1}{2} \end{cases}$$

$$g(x) = [\![x]\!]$$

$$h(x) = 3[\![x + 2]\!] - 1$$

1. $f(3)$

2. $f\left(\frac{1}{2}\right)$

3. $f\left(\frac{1}{3}\right)$

4. $f\left(\frac{5}{2}\right)$

5. $g(3.2)$

6. $g(1.8)$

7. $g(-2.4)$

8. $g(-6.9)$

9. $h(1.8)$

10. $h(3.1)$

11. $h(-0.4)$

12. $h(-3.1)$

Graph the function.

13. $f(x) = \begin{cases} x + 3, & \text{if } x < \frac{1}{2} \\ 2x - 1, & \text{if } x \geq \frac{1}{2} \end{cases}$

14. $f(x) = \begin{cases} 2x, & \text{if } x < -2 \\ x - 2, & \text{if } -2 \leq x \leq 2 \\ -2x, & \text{if } x > 2 \end{cases}$

15. $f(x) = \begin{cases} x + 1, & \text{if } x < 1 \\ 3x - 1, & \text{if } x > 1 \end{cases}$

16. $f(x) = [\![x + 1]\!]$

17. $f(x) = 3[\![x - 2]\!]$

18. $f(x) = [\![4x]\!]$

19. $f(x) = [\![2x + 3]\!]$

20. $f(x) = 2[\![3x - 1]\!] + 4$

21. $f(x) = -[\![2x + 1]\!] - 3$

Engraving **In Exercises 22–24, use the following information.**
A gift shop sells pewter mugs for $35. They are currently running an engraving promotion. The first six letters are engraved free. Each additional letter costs $0.20.

22. Write a piecewise model that gives the price of the mug with x engraved letters.

23. Graph the function.

24. What is the price of a mug with the name Jamie Lynn Krane engraved?

25. *Commission Sales*   A company pays its employees a combination of salary and commission. An employee with sales less than $100,000 earns a $15,000 salary plus 3% commission. An employee with sales of $100,000 to $200,000 earns an $18,000 salary plus 4% commission. An employee who earns more than $200,000 in sales earns a $20,000 salary plus 5% commission. Write a piecewise model that gives the pay of an employee with x in annual sales.

26. *Absolute value*   Write the function $f(x) = |x|$ as a piecewise function.

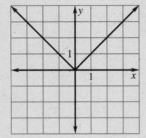

27. *Absolute value*   Write the function $f(x) = |x + 3| - 2$ as a piecewise function.

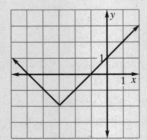

NAME _____ DATE _____

Reteaching with Practice

For use with pages 114–120

GOAL **Represent piecewise functions**

> ## VOCABULARY
>
> **Piecewise functions** are represented by a combination of equations, each corresponding to a part of the domain.
>
> A **step function** has a graph which resembles a set of stair steps. An example of a step function is the *greatest integer function*. This function is denoted by $g(x) = [\![x]\!]$, where for every real number x, $g(x)$ is the greatest integer less than or equal to x.

EXAMPLE 1 *Evaluating a Piecewise Function*

Evaluate $f(x)$ when (a) $x = -1$, (b) $x = 1$, and (c) $x = 3$.

$$f(x) = \begin{cases} 2x + 3, & \text{if } x < 0 \\ 2, & \text{if } 0 \le x < 2 \\ -x + 1, & \text{if } x \ge 2 \end{cases}$$

SOLUTION

a. $f(x) = 2x + 3$ Because $-1 < 0$, use first equation.

   $f(-1) = 2(-1) + 3 = 1$ Substitute -1 for x.

b. $f(x) = 2$ Because $0 \le 1 < 2$, use second equation.

   $f(1) = 2$ Substitute 1 for x.

c. $f(x) = -x + 1$ Because $3 \ge 2$, use third equation.

   $f(3) = -3 + 1 = -2$ Substitute 3 for x.

Exercises for Example 1

Evaluate the function for the given value of x.

$$f(x) = \begin{cases} x + 1, & \text{if } x > 1 \\ -x - 2, & \text{if } x \le 1 \end{cases} \qquad g(x) = \begin{cases} 3x + 2, & \text{if } x < 5 \\ -2x, & \text{if } x \ge 5 \end{cases}$$

1. $g(5)$ **2.** $f(0)$ **3.** $f(3)$ **4.** $g(-2)$

NAME _____ DATE _____

Reteaching with Practice

For use with pages 114–120

EXAMPLE 2 *Graphing a Piecewise Function*

Graph the function: $f(x) = \begin{cases} -x, & \text{if } x \le 3 \\ \frac{2}{3}x - 4, & \text{if } x > 3 \end{cases}$

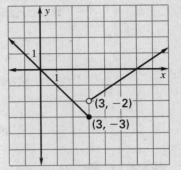

SOLUTION

To the right of $x = 3$, the graph is given by $y = \frac{2}{3}x - 4$. To the left of and including $x = 3$, the graph is given by $y = -x$.

The graph consists of two rays.

Exercises for Example 2

Graph the function.

5. $f(x) = \begin{cases} x + 2, & \text{if } x > 1 \\ -x + 2, & \text{if } x \le 1 \end{cases}$

6. $f(x) = \begin{cases} \frac{1}{2}x + 4, & \text{if } x < 2 \\ -2x + 9, & \text{if } x \ge 2 \end{cases}$

EXAMPLE 3 *Graphing a Step Function*

Graph the function. $f(x) = \begin{cases} -2, & \text{if } 0 \le x < 2 \\ -4, & \text{if } 2 \le x < 4 \\ -6, & \text{if } 4 \le x < 6 \end{cases}$

SOLUTION

The graph is composed of three line segments, because the function has three parts. The intervals of x tell you that each line segment is 2 units in length and begins with a solid dot and ends with an open dot.

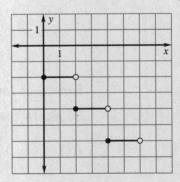

Exercises for Example 3

Graph the step function.

7. $f(x) = \begin{cases} 1, & \text{if } 0 < x \le 1 \\ 3, & \text{if } 1 < x \le 2 \\ 4, & \text{if } 2 < x \le 3 \\ 6, & \text{if } 3 < x \le 4 \end{cases}$

8. $f(x) = \begin{cases} -1, & \text{if } -2 \le x < 1 \\ -2, & \text{if } 1 \le x < 3 \\ -3, & \text{if } 3 \le x < 6 \\ -4, & \text{if } 6 \le x < 8 \end{cases}$

Lesson 2.7

NAME _____ DATE _____

Quick Catch-Up for Absent Students

For use with pages 114–121

The items checked below were covered in class on (date missed) _____

Lesson 2.7: Piecewise Functions

_____ **Goal 1:** Represent piecewise functions. (pp. 114–115)

Material Covered:

_____ Example 1: Evaluating a Piecewise Function

_____ Example 2: Graphing a Piecewise Function

_____ Example 3: Graphing a Step Function

_____ Example 4: Writing a Piecewise Function

Vocabulary:

piecewise function, p. 114   step function, p. 115

_____ **Goal 2:** Use piecewise functions to model real-life quantities. (p. 116)

Material Covered:

_____ Example 5: Using a Step Function

_____ Example 6: Using a Piecewise Function

Activity 2.7: Graphing Piecewise Functions (p. 121)

_____ **Goal:** Graph a piecewise function using a graphing calculator.

_____ Student Help: Keystroke Help

_____ Student Help: Look Back

_____ Other (specify) _____

_____

Homework and Additional Learning Support

_____ Textbook (specify) pp. 117–120 _____

_____

_____ *Reteaching with Practice* worksheet (specify exercises) _____

_____ *Personal Student Tutor* for Lesson 2.7

NAME _____ DATE _____

Interdisciplinary Application

For use with pages 114–120

Antiques

ECONOMICS   Antique dealers make their living through buying, selling, and shipping antiques to collectors all over the world.

Suppose you start an antique dealership that deals with African art. Collectors pay you to research, find pieces, and have the pieces shipped to them. In preparation for your business, you have gathered the following information from several shipping companies to determine their prices. Notice that some companies charge based on the weight of the package while others charge a flat rate.

Name of Shipper	Up to 10 lb	10–25 lb	Over 25 lb
Express Move	$1.25 per pound	$1.00 per pound	$0.75 per pound
Ship Rite	$11.25 flat rate	$18.00 flat rate	$25.00 flat rate
Packages Inc.	$2.00 per pound	An additional $1.50 for each pound over 10 pounds	An additional $1.10 for each pound over 25 pounds

1. Write a piecewise-defined function representing the cost of shipping x pounds with Express Move.

2. Write a piecewise-defined function representing the cost of shipping x pounds with Ship Rite.

3. Write a piecewise-defined function representing the cost of shipping x pounds with Packages Inc.

4. On the same set of axes, graph the functions that represent the costs for the three different companies.

5. Suppose you were shipping a 17-pound statue from Ghana and had the choice of the three companies here. Which would you pick? How much would the shipping cost be?

6. Suppose you knew that all of your packages are between 10 and 25 pounds. Which company (or companies) provide the least expensive shipping in this range?

7. When is Packages Inc. the cheapest carrier?

Challenge: Skills and Applications

For use with pages 114–120

1. Let $f(x)$ be defined as follows:

$$f(x) = \begin{cases} 2 \text{ if } x \text{ is an integer} \\ 1 \text{ if } x \text{ is not an integer} \end{cases}$$

   a. Find each of the following: $f(0), f\left(\frac{1}{3}\right), f\left(\frac{1}{2}\right), f\left(\sqrt{2}\right), f(2), f(\pi)$

   b. Sketch the graph of $y = f(x)$.

2. What is perhaps the ultimate piecewise-defined function was proposed by the German mathematician Gustav Peter Lejeune Dirichlet (1805–1859). It is defined as follows:

$$f(x) = \begin{cases} 1 \text{ if } x \text{ is a rational number} \\ 0 \text{ if } x \text{ is an irrational number} \end{cases}$$

   a. Find each of the following values: $f(2), f\left(\frac{1}{2}\right), f\left(\sqrt{2}\right), f\left(\frac{5}{4}\right), f(\pi)$.

   b. Explain how you know that Dirichlet's function is actually a function, and not just a relation. Could a computer draw an accurate graph of this function?

3. Suppose m and k are numbers and let $g(x)$ be defined piecewise as follows:

$$g(x) = \begin{cases} \frac{1}{2}x + k \text{ if } x < 3 \\ mx + 2 \text{ if } x \geq 3 \end{cases}$$

   a. Suppose $m = \frac{3}{2}$. What must the value of k be in order for the graph of $y = g(x)$ to be connected?

   b. Suppose $k = 4$. What must the value of m be in order for the graph to be connected?

4. Let $f(x)$ be defined for all nonnegative real numbers as follows:

$$f(x) = \begin{cases} 1 \text{ if } x = 0 \\ f(0)x + f(0) \text{ if } 0 < x \leq 1 \\ f(1)(x - 1) + f(1) \text{ if } 1 < x \leq 2 \\ f(2)(x - 2) + f(2) \text{ if } 2 < x \leq 3 \\ \dots \end{cases}$$

   a. Find $f(1), f(2), f(3)$, and $f(4)$.

   b. Sketch the graph of $f(x)$ for $0 \leq x \leq 4$.

TEACHER'S NAME _____ CLASS _____ ROOM _____ DATE _____

Lesson Plan

1-day lesson (See *Pacing the Chapter,* TE pages 64C–64D) **For use with pages 122–128**

GOALS 1. **Represent absolute value functions.**
2. **Use absolute value functions to model real-life situations.**

State/Local Objectives _____

_____

✓ Check the items you wish to use for this lesson.

STARTING OPTIONS
____ Homework Check: TE page 117; Answer Transparencies
____ Warm-Up or Daily Homework Quiz: TE pages 122 and 120, CRB page 106, or Transparencies

TEACHING OPTIONS
____ Motivating the Lesson: TE page 123
____ Lesson Opener (Graphing Calculator): CRB page 107 or Transparencies
____ Graphing Calculator Activity with Keystrokes: CRB pages 108–109
____ Examples 1–4: SE pages 123–124
____ Extra Examples: TE pages 123–124 or Transparencies
____ Closure Question: TE page 124
____ Guided Practice Exercises: SE page 125

APPLY/HOMEWORK
Homework Assignment
____ Basic 12–17, 18–38 even, 40, 41, 49, 50, 57–65 odd; Quiz 3: 1–14
____ Average 12–17, 18–38 even, 40–45, 49, 50, 57–65 odd; Quiz 3: 1–14
____ Advanced 12–17, 18–38 even, 40–45, 49–55, 57–65 odd; Quiz 3: 1–14

Reteaching the Lesson
____ Practice Masters: CRB pages 110–112 (Level A, Level B, Level C)
____ Reteaching with Practice: CRB pages 113–114 or Practice Workbook with Examples
____ Personal Student Tutor

Extending the Lesson
____ Applications (Real-Life): CRB page 116
____ Challenge: SE page 127; CRB page 117 or Internet

ASSESSMENT OPTIONS
____ Checkpoint Exercises: TE pages 123–124 or Transparencies
____ Daily Homework Quiz (2.8): TE page 128 or Transparencies
____ Standardized Test Practice: SE page 127; TE page 128; STP Workbook; Transparencies
____ Quiz (2.6–2.8): SE page 128

Notes _____

_____

_____

TEACHER'S NAME _____ CLASS _____ ROOM _____ DATE _____

Lesson Plan for Block Scheduling

Half-day lesson (See *Pacing the Chapter*, TE pages 64C–64D)   For use with pages 122–128

GOALS   1. **Represent absolute value functions.**
            2. **Use absolute value functions to model real-life situations.**

State/Local Objectives _____

_____

_____

CHAPTER PACING GUIDE	
Day	**Lesson**
1	2.1 (all); 2.2 (all)
2	2.3 (all)
3	2.4 (all)
4	2.5 (all); 2.6 (all)
5	2.7 (all); **2.8 (all)**
6	Review/Assess Ch. 2

✓ **Check the items you wish to use for this lesson.**

STARTING OPTIONS

_____ Homework Check: TE page 117; Answer Transparencies

_____ Warm-Up or Daily Homework Quiz: TE pages 122 and 120,
         CRB page 106, or Transparencies

TEACHING OPTIONS

_____ Motivating the Lesson: TE page 123

_____ Lesson Opener (Graphing Calculator): CRB page 107 or Transparencies

_____ Graphing Calculator Activity with Keystrokes: CRB pages 108–109

_____ Examples 1–4: SE pages 123–124

_____ Extra Examples: TE pages 123–124 or Transparencies

_____ Closure Question: TE page 124

_____ Guided Practice Exercises: SE page 125

APPLY/HOMEWORK

Homework Assignment (See also the assignment for Lesson 2.7.)

_____ Block Schedule: 12–17, 18–38 even, 40–45, 49, 50, 57–65 odd; Quiz 3: 1–14

Reteaching the Lesson

_____ Practice Masters: CRB pages 110–112 (Level A, Level B, Level C)

_____ Reteaching with Practice: CRB pages 113–114 or Practice Workbook with Examples

_____ Personal Student Tutor

Extending the Lesson

_____ Applications (Real Life): CRB page 116

_____ Challenge: SE page 127; CRB page 117 or Internet

ASSESSMENT OPTIONS

_____ Checkpoint Exercises: TE pages 123–124 or Transparencies

_____ Daily Homework Quiz (2.8): TE page 128 or Transparencies

_____ Standardized Test Practice: SE page 127; TE page 128; STP Workbook; Transparencies

_____ Quiz (2.6–2.8): SE page 128

Notes _____

_____

_____

NAME _____ DATE _____

WARM-UP EXERCISES

For use before Lesson 2.8, pages 122–128

Evaluate the expression for $x = -6$.

1. $|x|$

2. $-|x - 3|$

3. $|1 - x| + 4$

4. $2|x - 1.5| + 0.5$

5. $-3|x + 4| - 1$

· ·

DAILY HOMEWORK QUIZ

For use after Lesson 2.7, pages 114–121

1. Evaluate $f(x) = \begin{cases} x, & \text{if } x > 3 \\ 2x - 1, & \text{if } x \le 3 \end{cases}$ for $f(4)$ and $f(0)$.

2. Graph $f(x) = \begin{cases} -x, & \text{if } x \le 1 \\ \frac{3}{2}x - \frac{3}{2}, & \text{if } x > 1 \end{cases}$.

3. Write equations for the step function whose graph is shown.

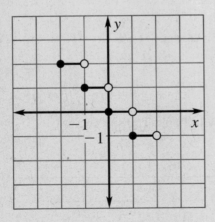

Graphing Calculator Activity Opener

For use with pages 122–128

SET UP: Work with a partner.
YOU WILL NEED: • Graphing calculator

Choose one student to be the Grapher and the other to be the
Guesser.

1. Grapher: Choose a function of the form $y = a|x|$ and graph it on
   the graphing calculator as y_1. Do not let the Guesser see the
   function, only the graph.

   Guesser: Look at the graph and guess the function.

   Grapher: Graph the guess as y_2.

   Guesser: If your guess is not correct, guess again.

   Grapher: Change y_2 to the new guess and graph again.

   Continue until the graphs of the two functions match.

2. Change roles and repeat the activity. Make a equal to a negative
   number or to a number between 0 and 1. Adjust the viewing
   window as needed.

3. Repeat the activity two more times using functions of the form
   $y = |x| + k$, alternating roles.

4. Repeat the activity two more times using functions of the form
   $y = |x - h|$, alternating roles.

Lesson 2.8

Graphing Calculator Activity

For use with pages 122–128

GOAL **To graph equations of absolute value functions**

Recall that $|x| = \begin{cases} x, \text{ if } x \geq 0 \\ 0, \text{ if } x = 0 \\ -x, \text{ if } x < 0 \end{cases}$

Activity

❶ Use the above definition to graph $y = |x|$.

❷ Use a graphing calculator to check your work in Step 1.

❸ Use a graphing calculator to graph each of these functions in the same viewing window.

$y = |x| + 3$ $y = |x| + 5$ $y = |x| + 7$

❸ Repeat Step 3 for these functions.

$y = |x| - 2$ $y = |x| - 4$ $y = |x| - 6$

❸ Describe the effect of c on the graph of $y = |x| + c$.

Exercises

1. Use the following phrases to describe how the graph of each function in parts (a)-(d) is related to the graph of $y = |x|$.

- shifted up 1 unit
- shifted up 2 units
- shifted down 1 unit
- shifted down 3 units

(a) $y = |x| + 1$ (b) $y = |x| + 2$ (c) $y = |x| - 3$ (d) $y = |x| - 1$

2. Use a graphing calculator to graph each of these functions in the same viewing window.

$y = |x|$ $y = |x - 5|$ $y = |x + 4|$

3. Described the effect of h on the graph of $y = |x - h|$.

Graphing Calculator Activity

For use with pages 122–128

TI-82

Step 2:

[Y=] [2nd] [ABS] [X,T,θ] [ZOOM] 6

Step 3

[Y=] [2nd] [ABS] [X,T,θ] [+] 3 [ENTER] [2nd]
[ABS] [X,T,θ] [+] 5 [ENTER] [2nd] [ABS] [X,T,θ]
[+] 7 [ENTER] [GRAPH]

Step 4:

[Y=] [2nd] [ABS] [X,T,θ] [–] 2 [ENTER] [2nd]
[ABS] [X,T,θ] [–] 4 [ENTER] [2nd] [ABS] [X,T,θ]
[–] 6 [ENTER] [GRAPH]

TI-83

Step 2:

[Y=] [MATH] [►] 1 [X,T,θ,n] [ )] [ZOOM] 6

Step 3

[Y=] [MATH] [►] 1 [X,T,θ,n] [ )] 3 [ENTER] [MATH]
[►] 1 [X,T,θ,n] [ )] [+] 5 [ENTER] [MATH] [►]
1 [X,T,θ,n] [ )] [+] 7 [ENTER] [GRAPH]

Step 4:

[Y=] [MATH] [►] 1 [X,T,θ,n] [ )] [–] 2 [ENTER]
[MATH] [►] 1 [X,T,θ,n] [ )] [–] 4 [ENTER] [MATH]
[►] 1 [X,T,θ,n] [ )] [–] 6 [ENTER] [GRAPH]

SHARP EL-9600c

Step 2:

[Y=] [MATH] [B] 1 [X/θ/T/n] [ZOOM] 6

Step 3:

[Y=] [MATH] [B] 1 [X/θ/T/n] [►] [+] 3 [ENTER]
[MATH] [B] 1 [X/θ/T/n] [►] [+] 5 [ENTER] [MATH]
[B] 1 [X/θ/T/n] [►] [+] 7 [ENTER] [GRAPH]

Step 4:

[Y=] [MATH] [B] 1 [X/θ/T/n] [►] [–] 2 [ENTER]
[MATH] [B] 1 [X/θ/T/n] [►] [–] 4 [ENTER] [MATH]
[B] 1 [X/θ/T/n] [►] [–] 6 [ENTER] [GRAPH]

CASIO CFX-9850GA PLUS

Step 2:

From the main menu, choose GRAPH.

[OPTN] [F5] [F1] [X,θ,T] [EXE] [SHIFT] [F3] [F3]
[EXIT] [F6]

Step 3:

From the main menu, choose GRAPH.

[OPTN] [F5] [F1] [X,θ,T] [+] 3 [EXE] [OPTN] [F5]
[F1] [X,θ,T] [+] 5 [EXE] [OPTN] [F5] [F1] [X,θ,T]
[+] 7 [EXE] [F6]

Step 4:

From the main menu, choose GRAPH.

[OPTN] [F5] [F1] [X,θ,T] [–] 2 [EXE] [OPTN] [F5]
[F1] [X,θ,T] [–] 4 [EXE] [OPTN] [F5] [F1] [X,θ,T]
[–] 6 [EXE] [F6]

Lesson 2.8

Practice A

For use with pages 122–128

Match the function with its graph.

1. $f(x) = |x + 4|$

2. $f(x) = |x - 4|$

3. $f(x) = |x| + 4$

4. $f(x) = |x| - 4$

5. $f(x) = 4|x|$

6. $f(x) = \frac{1}{4}|x|$

A.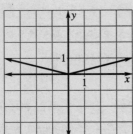

B.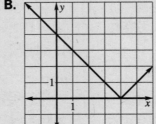

C.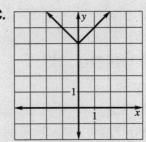

D.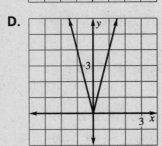

E.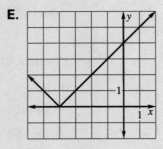

F.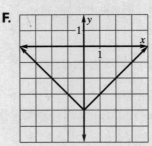

Tell whether the graph of the function opens up or down.

7. $y = -3|x|$

8. $y = 3|x + 1|$

9. $y = |x + 1| - 10$

10. $y = 4|x - 1| + 3$

11. $y = -2|x + 1| + 7$

12. $y = -|x - 2| + 4$

Identify the vertex of the graph of the given function.

13. $y = 2|x| - 3$

14. $y = |x - 1| + 2$

15. $y = |x + 3| - 5$

16. $y = |x - 7| - 2$

17. $y = 2|x + 1| + 9$

18. $y = -5|x + 3|$

Tell whether the graph of the function is *wider, narrower,* or the same width as the graph of $y = |x|$.

19. $y = |x - 8|$

20. $y = 2|x - 1|$

21. $y = \frac{1}{2}|x + 3| - 2$

22. $y = -3|x + 1| + 7$

23. $y = -\frac{2}{3}|x - 6| + 3$

24. $y = \frac{9}{10}|x| + 13$

Swimwear **In Exercises 25 and 26, use the following information.**

A sporting goods store sells swimming suits year round. The number of suits sold can be modeled by the function $S = -90|t - 6| + 540$, where t is the time in months and S is the sales in dollars.

25. Graph the function for $0 \le t \le 12$.

26. What is the maximum sales in one month? In what month is the maximum reached?

NAME _____ DATE _____

Practice B

For use with pages 122–128

Tell whether the graph of the function opens up or down.

1. $y = |x + 3| - 5$

2. $y = -4|x - 1| + 6$

3. $y = \frac{2}{3}|x - 2| + 9$

Identify the vertex of the graph of the given function.

4. $y = 2|x + 13| - 6$

5. $y = -3|x - 4| - 7$

6. $y = \frac{1}{5}|x + 2| + 11$

Tell whether the graph is *wider, narrower,* or the *same width* as the graph of $y = |x|$.

7. $y = \frac{3}{5}|x - 3| + 7$

8. $y = -8|x + 9| - 12$

9. $y = -\frac{5}{2}|x - 1| - 3$

Graph the function.

10. $y = |x| - 4$

11. $y = |x - 4|$

12. $y = |x + 2| - 3$

13. $y = |x + 1| + 3$

14. $y = 2|x - 3|$

15. $y = -|x + 5|$

16. $y = |x - 4| + 5$

17. $y = 3|x - 1| - 2$

18. $y = -2|x + 7| - 4$

19. $y = \frac{1}{2}|x| - 2$

20. $y = \frac{2}{3}|x + 2| + 1$

21. $y = -\frac{1}{2}|x - 1| + 2$

Write an equation of the graph shown.

22.

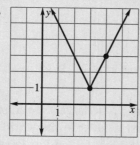

23.

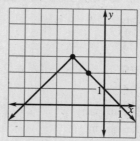

24.

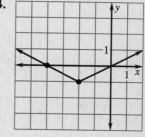

A-Frame Home   **In Exercises 25 and 26, use the following information.**

The roof line of an A-frame home follows the path given by $y = -\frac{11}{6}|x| + 22$. Each unit on the coordinate plane represents one foot.

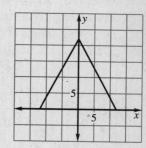

25. Find the vertex of the graph.

26. What does the y-value of the vertex tell us about the home?

Fine Dining   **In Exercises 27 and 28, use the following information.**

An exclusive restaurant is open from 3:00 P.M. to 10:00 P.M. Each evening, the number of people served S increases steadily and then decreases according to the model $S = -30|t - 6.5| + 105$ where $t = 0$ represents 12:00 P.M.

27. Graph the function.

28. Find the vertex of the graph. Explain what each coordinate of the vertex represents.

Algebra 2
Chapter 2 Resource Book

Lesson 2.8

NAME _____ DATE _____

Practice C

For use with pages 122–128

Tell whether the graph of the function opens up or down.

1. $y = -\dfrac{1}{3}|x + 2| + 4$ **2.** $y = 3 + \dfrac{1}{2}|x + 1|$ **3.** $y = 4 - 2|x + 3|$

Identify the vertex of the graph of the given function.

4. $y = 3|x - 2| + 5$ **5.** $y = \dfrac{1}{3}\left|x - \dfrac{3}{8}\right| - 1$ **6.** $y = 6 + \dfrac{4}{5}\left|x - \dfrac{2}{3}\right|$

Tell whether the graph is *wider, narrower,* or the *same width* as the graph of $y = |x|$.

7. $y = -\dfrac{2}{3}|x + 1| - 4$ **8.** $y = \dfrac{5}{6}|x - 3| + 1$ **9.** $y = 4 - \dfrac{7}{6}|x + 3|$

Graph the function.

10. $y = 2|x + 1| - 4$ **11.** $y = -3|x - 3| + 2$ **12.** $y = -4 + 5|x + 2|$

13. $y = \dfrac{1}{2}|x - 3| + 1$ **14.** $y = -\dfrac{1}{3}|x + 2| + 3$ **15.** $y = 2\left|x - \dfrac{1}{2}\right| + 3$

16. $y = -\left|x + \dfrac{2}{3}\right| - 1$ **17.** $y = 2.5|x - 1.3| - 2.4$ **18.** $y = -1.8|x + 2.2| + 1.6$

Graph the function by making a table and plotting points. Then write a function of the form $y = a|x - h| + k$ that has the same graph.

19. $y = |2x|$ **20.** $y = |-3x|$ **21.** $y = |2x + 6|$

22. $y = |-5x + 20|$ **23.** $y = |4x| + 2$ **24.** $y = |-2x| - 3$

25. $y = |-2x - 8| + 1$ **26.** $y = |3x - 9| + 2$ **27.** $y = 2|2x + 10| + 1$

28. *Company's Profit* The profit for a company from 1988 to 1998 is modeled by the graph. The profit is measured in thousands of dollars and $t = 0$ corresponds to 1988. Write a piecewise function that represents the profit.

29. *Pyramids of Egypt* The largest pyramid included in the first wonder of the world is Khufu. It stands 450 feet tall and its base is 755 feet long. Imagine that a coordinate plane is placed over a side of the pyramid. In the coordinate plane, each unit represents one foot and the origin is at the center of the pyramid's base. Write an absolute value function for the outline of the pyramid.

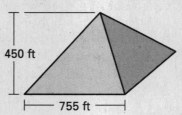

450 ft

755 ft

NAME _____ DATE _____

Reteaching with Practice

For use with pages 122–128

GOAL Represent absolute value functions and use absolute value functions to model real-life situations

EXAMPLE 1 *Graphing an Absolute Value Function*

Graph $y = 3|x - 2| - 4$.

SOLUTION

First plot the vertex at $(2, -4)$. Then plot another point, such as $(1, -1)$. Use symmetry to plot a third point, $(3, -1)$. Connect these three points with a V-shaped graph. Notice that $a = 3 > 0$ and $|a| > 1$, so the graph opens up and is narrower than $y = |x|$.

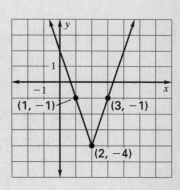

Exercises for Example 1

Graph the function.

1. $y = -2|x + 1| + 3$ **2.** $y = |x - 3| + 4$ **3.** $y = |x| + 5$

4. $y = 5|x - 2|$ **5.** $y = -|x - 1| - 3$ **6.** $y = -|x| - 2$

NAME _____ DATE _____

Reteaching with Practice

For use with pages 122–128

EXAMPLE 2 *Writing an Absolute Value Function*

Write an equation of the graph shown.

SOLUTION

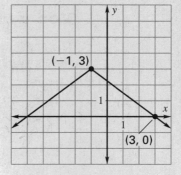

The vertex is $(-1, 3)$, so the equation has the form
$y = a|x + 1| + 3$.

To find a, substitute the coordinates of the point $(3, 0)$
into the equation and solve.

$y = a|x + 1| + 3$	Write equation.
$0 = a|3 + 1| + 3$	Substitute 0 for y and 3 for x.
$0 = 4a + 3$	Simplify.
$-3 = 4a$	Subtract 3 from each side.
$-\frac{3}{4} = a$	Solve for a.

An equation of the graph is $y = -\frac{3}{4}|x + 1| + 3$.

✓ CHECK Notice the graph opens down since $a = -\frac{3}{4} < 0$, and it is
wider than the graph of $y = |x|$ since $|a| = \left|-\frac{3}{4}\right| = \frac{3}{4} < 1$.

Exercises for Example 2

Write an equation of the graph shown.

7.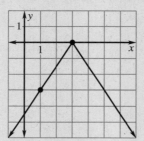

8.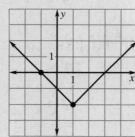

9.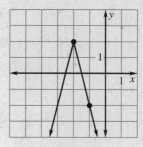

NAME _____ DATE _____

Quick Catch-Up for Absent Students

For use with pages 122–128

The items checked below were covered in class on (date missed) _____

Lesson 2.8: Absolute Value Functions

_____ **Goal 1:** Represent absolute value functions. (pp. 122–123)

Material Covered:

_____ Activity: Graphs of Absolute Value Functions

_____ Student Help: Skills Review

_____ Example 1: Graphing an Absolute Value Function

_____ Example 2: Writing an Absolute Value Function

Vocabulary:

vertex of an absolute value graph, p. 122

_____ **Goal 2:** Use absolute value functions to model real-life situations. (p. 124)

Material Covered:

_____ Example 3: Interpreting an Absolute Value Function

_____ Example 4: Interpreting an Absolute Value Graph

_____ Other (specify) _____

_____

Homework and Additional Learning Support

_____ Textbook (specify) <u>pp. 125–128</u> _____

_____

_____ *Reteaching with Practice* worksheet (specify exercises) _____

_____ *Personal Student Tutor* for Lesson 2.8

NAME _____ DATE _____

Real–Life Application: When Will I Ever Use This?

For use with pages 122–128

City Planning

Many small towns in the United States do not have a large enough population to support a rescue squad or paramedic unit by themselves. Frequently they must depend on larger nearby towns to serve their emergency medical needs. However, these small towns often have a small group of people to serve as *first responders,* trained medical personnel that will report to an emergency call to stabilize the patient while waiting for the paramedics from the larger town.

Plainview, North Carolina, is such a town. It is situated 15 miles outside of Dunn, North Carolina, and has most of its residences along a single, long highway. The Plainview city council has approved money to build a small rescue station along the highway to serve as a headquarters for the Plainview first responders. Your job is to determine where along the highway the rescue station should be built.

Assume that the highway is a straight line. The point on the highway where the town limits start is marked as zero. The town has residences at points 0.5, 1, 1.2, 1.3, 1.8, 2, and 2.5 miles from point zero. Answer the following questions to determine the location where the rescue station should be built.

1. Let x represent the unknown position of the rescue station. Write an absolute value expression for the distance between the station and the house located at the 1-mile marker.

2. Write absolute value expressions for the distance between the station and the rest of the houses in Plainview.

3. Let y represent the total distance between the station and all of the houses. Write an algebraic equation for y.

4. Use a graphing calculator to graph the equation in Exercise 3.

5. Suppose a Plainview town council member would only approve the funding if the station could be located within 0.5 mile of each house. Is this possible?

6. Use your graph to find the position along the highway that minimizes the total distance between the station and all of the houses.

Challenge: Skills and Applications

For use with pages 122–128

Graph each inequality.

1. $2|x - 1| < y + 3$

2. $y \le -\frac{1}{2}|x + 1| + 4$

Graph each absolute value equation.

3. $|x + y| = 5$

4. $|x| + |y| = 6$

5. The graph shown at the right is that of a typical absolute value function $y = a|x - h| + k$, with $h > 0$ and $k > 0$. Find, in terms of *a, k,* and *h,* the area of the region between the graph and the *x*-axis, and between the *y*-axis and the line $x = 2h$.

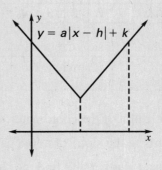

6. Two friends, Jin Chang and Juanita Alvarez, decide to meet for dinner at a restaurant. Each will try to arrive some time after 5:00 P.M., and each will wait 10 minutes for the other, if she is not already there.

   a. Let $x =$ Linda's arrival time and let $y =$ Juanita's arrival time, in minutes after 5:00 P.M. Write an absolute value inequality that expresses that they will arrive at times that differ by no more than 10 minutes.

   b. Graph the absolute value inequality you wrote in part (a).

7. Imagine that you are "standing on" the *y*-axis, with your feet at point *A* and your head at point *C*. Your eyes are at point *B*. You would like to put a mirror along the *x*-axis that would allow you to see the reflection of your whole body.

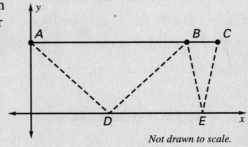

Not drawn to scale.

   a. Suppose points *A, B,* and *C* have coordinates (0, 40), (64, 40), and (68, 40). Find an absolute value function that models the path of light from your feet to your eyes: *ADB,* and another absolute value function that models the path of light from the top of your head to you eyes: *CEB.*

   b. What is the minimum length the mirror can have? How does this relate to your height?

Lesson 2.8

NAME _____ DATE _____

Chapter Review Games and Activities

For use after Chapter 2

1. Each member of the class should measure their height and the length of their foot in inches.

   Let x = height and y = length of foot.

2. Complete a table such as the one below for each member of the class.

Inches	1	2	3	4	5	6	7	8	9	. . .	# in class
x = height											
y = foot length											

3. Make a plot of the coordinates of the data.

4. Determine the slope of a line passing through any two given points.

5. Choose a partner and find the slope of a line passing through your coordinates and your partner's coordinates.

6. Determine the equation of the line passing through your coordinates and the coordinates of your partner.

   a. Give the slope and y-intercept of both of your lines.

   b. Write your equations in slope-intercept form.

   c. Write your equations in point-slope form.

7. Graph both lines in the Cartesian plane.

8. Approximate a line of best fit for your data.

9. Compare your line to that of the rest of the class.

10. Plot all class data on the same graph.

11. Sketch a line of best fit.

12. Choose two points to find a line of best fit.

13. Ask your teacher or a student from another class their height.

14. See how closely you can predict the length of their foot with the class's line of best fit.

Algebra 2
Chapter 2 Resource Book

NAME _____ DATE _____

Chapter Test A

For use after Chapter 2

Graph the relation. Then tell whether the relation is a function.

1.

x	0	1	2	−1	−2
y	1	2	3	0	1

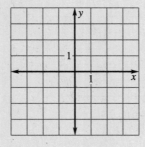

2.

x	3	4	5	3	0	−1
y	3	4	5	6	0	−1

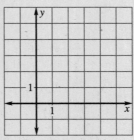

Evaluate the function for the given value of x.

3. $f(x) = x + 3$ when $x = -3$ 4. $f(x) = x^2 - 3x$ when $x = 3$

5. $f(x) = |x - 2|$ when $x = 5$

Graph the equation.

6. $x = -1$ 7. $y = \frac{2}{3}x - 2$

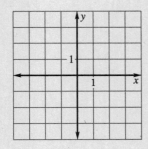   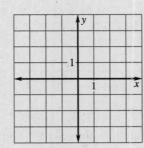

Write an equation of the line that has the given properties.

8. slope: $\frac{1}{2}$,   9. slope: 2,   10. points: (2, 1),
   y-intercept: 0   point: (1, 3)   (3, 2)

11. Write an equation of the line that passes through (4, 3) and is parallel to the line $y = x + 1$.

12. Write an equation of the line that passes through (−2, 1) and is perpendicular to the line $y = \frac{1}{2}x + 1$.

Answers

1. Use grid at left.

_____

2. Use grid at left.

_____

3. _____

4. _____

5. _____

6. Use grid at left.

7. Use grid at left.

8. _____

9. _____

10. _____

11. _____

12. _____

Review and Assess

Graph the inequality in a coordinate plane.

13. $y \geq 2$

14. $y \geq 2x + 1$

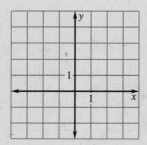

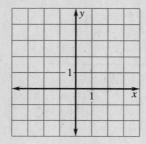

Graph the function.

15. $f(x) = \begin{cases} 0, \text{ if } x > 0 \\ 2, \text{ if } x \leq 0 \end{cases}$

16. $f(x) = \begin{cases} x + 2, & \text{ if } x \geq 0 \\ x - 2, & \text{ if } x < 0 \end{cases}$

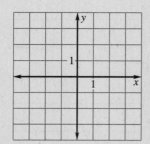

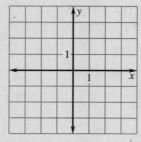

17. $f(x) = |x| + 1$

18. $f(x) = \frac{1}{2}|x - 4|$

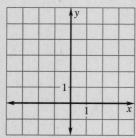

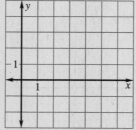

19. *Ticket Prices* Student tickets for a football game cost $2 each. Adult tickets cost $4 each. Ticket sales at last week's game totaled $2800. Write a model that shows the different numbers of students and adults who could have attended the game.

13.	Use grid at left.
14.	Use grid at left.
15.	Use grid at left.
16.	Use grid at left.
17.	Use grid at left.
18.	Use grid at left.
19.	

Review and Assess

NAME _____ DATE _____

Chapter Test B

For use after Chapter 2

Graph the relation. Then tell whether the relation is a function.

1.

x	−3	−1	0	2	3
y	−2	0	1	2	3

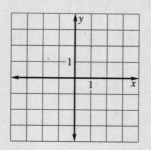

2.

x	3	−3	4	0	3	2
y	3	−2	1	1	4	2

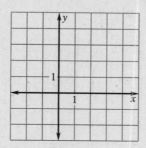

Evaluate the function for the given value of x.

3. $f(x) = 25 - 2x$ when $x = 4$

4. $f(x) = |x - 5|$ when $x = 2$

5. $f(x) = x^2 + 5x + 1$ when $x = 1$

Graph the equation.

6. $y = \frac{1}{2}x + 3$

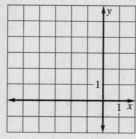

7. $y = 2$

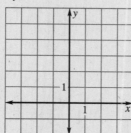

8. $4x - y = 8$

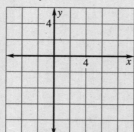

9. $x = -2$

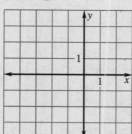

Write an equation of the line that has the given properties.

10. slope: $\frac{1}{2}$,
y-intercept: 2

11. slope: 1
point: $(2, -3)$

12. points:
$(3, 4), (-1, 0)$

Answers

1. Use grid at left. _____

_____

2. Use grid at left. _____

_____

3. _____

4. _____

5. _____

6. Use grid at left.

7. Use grid at left.

8. Use grid at left.

9. Use grid at left.

10. _____

11. _____

12. _____

Review and Assess

13. Write an equation of the line that passes through $(1, 6)$ and is parallel to the line $x - y = 4$.

14. Write an equation of the line that passes through $(-2, 3)$ and is perpendicular to the line $y = 2x + 1$.

Graph the inequality in a coordinate plane.

15. $y \geq 1$

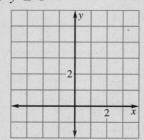

16. $y > 2x + 1$

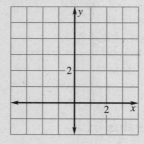

17. $x - 2y \leq 0$

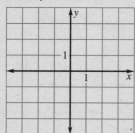

18. $x \geq -2$

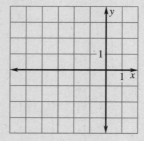

Graph the function.

19. $f(x) = \begin{cases} 1, \text{ if } x > 0 \\ -1, \text{ if } x < 0 \end{cases}$

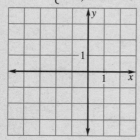

20. $f(x) = 4|x| - 3$

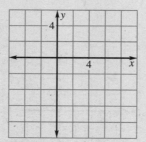

21. *Profit*   The sophomore class needs to raise money. They sell boxes of holiday cards at a profit of $2 per box. How many boxes must they sell to make a profit of at least $600? Express your answer as an inequality.

22. *Roofs*   A roof rises 3 units for every 4 units of horizontal run.
(a) What is the slope of the roof?
(b) If the roof is 15 feet high, how long is it?

13. _____

14. _____

15. _Use grid at left._

16. _Use grid at left._

17. _Use grid at left._

18. _Use grid at left._

19. _Use grid at left._

20. _Use grid at left._

21. _____

22. (a) _____

(b) _____

Review and Assess

Chapter Test C

For use after Chapter 2

Graph the relation. Then tell whether the relation is a function.

1.

x	1	2	3	4	5
y	2	2	2	2	2

2.

x	−3	−2	−1	0	1
y	3	−3	3	−3	3

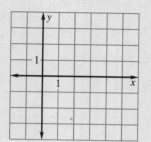

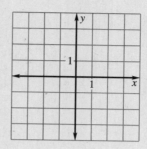

Evaluate the function for the given value of x.

3. $f(x) = |x|$ when $x = -5$

4. $f(x) = x^2 - x - 1$ when $x = 1$

5. $f(x) = 2(x^2 - 4) + 1$ when $x = 0$

Graph the equation.

6. $y = 0$

7. $y = \frac{2}{3}x + 1$

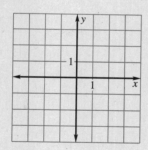

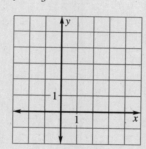

Write an equation of the line that has the given properties.

8. slope: $\frac{3}{4}$,

   y-intercept: -2

9. slope: $-\frac{1}{2}$,

   point: $(-4, -3)$

10. points: $(0, 5)$,

    $(-5, 0)$

11. Write an equation of the line that passes through $(5, 5)$ and is parallel to the line $2x - 2y = -3$.

12. Write an equation of the line that passes through $(-2, -4)$ and is perpendicular to the line $y = \frac{1}{2}x + 3$.

Answers

1. Use grid at left.

   _____

2. Use grid at left.

   _____

3. _____

4. _____

5. _____

6. Use grid at left.

7. Use grid at left.

8. _____

9. _____

10. _____

11. _____

12. _____

Review and Assess

Graph the inequality in a coordinate plane.

13. $x \geq -3$

14. $x > 2y + 1$

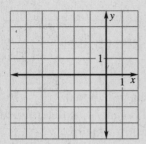

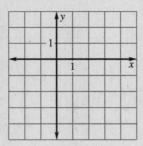

13.	Use grid at left.
14.	Use grid at left.
15.	Use grid at left.
16.	Use grid at left.
17.	Use grid at left.
18.	Use grid at left.
19.	

Graph the function.

15. $f(x) = \begin{cases} x, & \text{if } -2 < x < 2 \\ 2x, & \text{if } x < -2 \\ 3x, & \text{if } x > 2 \end{cases}$

16. $f(x) = \begin{cases} \frac{1}{2}x - 2, & \text{if } x \geq 0 \\ \frac{3}{4}x + 3, & \text{if } x < 0 \end{cases}$

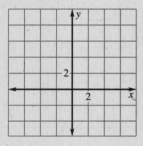

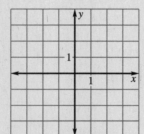

17. $f(x) = \frac{1}{2}|x| + 2$

18. $f(x) = \begin{cases} |x + 1|, & \text{if } x > 0 \\ |x - 1|, & \text{if } x < 0 \end{cases}$

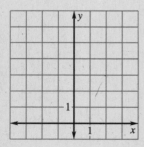

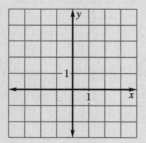

19. *Car Wash*   A local car wash charges $8 per wash and $10 per wash and wax. At the end of a certain day, the total sales were $3100. Write a model that shows the different numbers of the two types of car washes. Then find the number of wash and waxes there were if 200 were washes only.

NAME _____ DATE _____

SAT/ACT Chapter Test

For use after Chapter 2

1. The graph of $x + 2y = -5$ contains which point?

   A $(0, 0)$ **B** $(2, -1)$

   C $(-2, -1)$ **D** $(-1, -2)$

2. The slope of a line in the form $y = mx + b$ is

   A y.   **B** m.

   C x.   **D** b.

3. The solution of $|x + 2| = 4$ is

   A $2, -2$.   **B** $0, 2$.

   C $2, -4$.   **D** $2, -6$.

4. The inequaltity $-4 < x < 4$

   A is undefined.   **B** has no solution.

   C includes 0.   **D** includes 4.

5. The points $(1, 2)$, $(2, 3)$, $(3, 4)$, and $(4, 5)$ lie on the line with equation

   A $-x + y = 1$.   **B** $x - y = 1$.

   C $x + y = 1$.   **D** $x + y = -1$.

6. Write an equation of the line that has a slope of $\frac{1}{2}$ and a y-intercept of -2.

   A $y = -\frac{1}{2}x - 2$ **B** $y = \frac{1}{2}x + 2$

   C $y = \frac{1}{2}x - 2$ **D** $y = -\frac{1}{2}x + 2$

In Questions 7–10, choose the statement below that is true about the given numbers.

   A The number in column A is greater.

   B The number in column B is greater.

   C The two numbers are equal.

   D The relationship cannot be determined from the given information.

7.

Column A	Column B
$|x + 2|$	$|y + 3|$

A   **B**   **C**   **D**

8.

Column A	Column B
The slope of $y = 3x + 3$	The slope of $y = -3x + 3$

A   **B**   **C**   **D**

9.

Column A	Column B
$f(x) = 10 - x$ when $x = 11$	$f(x) = x - 10$ when $x = 11$

A   **B**   **C**   **D**

10.

Column A	Column B
The y-intercept of $x + y = -5$	The y-intercept of $x - y = -5$

A   **B**   **C**   **D**

Alternative Assessment and Math Journal

For use after Chapter 2

JOURNAL   **1.** Identifying a function is an important skill in mathematics. Explain how to determine whether parts (a) through (c) are functions.

  a. a set of ordered pairs

  b. a graph

  c. an equation

  d. Give an example for each of the above which is a function.

MULTI-STEP   **2.** Scores on a recent physics test were a little low. The teacher creates a
PROBLEM   linear function to convert the original scores into adjusted scores. An original score of 66 is transformed into an adjusted score of 76. The original high score of 87 was converted to a high score of 90 after the adjustment.

  a. Write the given information about the score adjustments as ordered pairs.

  b. Find the linear equation which will convert any original score into an adjusted score.

  c. If a student's original score was 78, find the adjusted score.

  d. What original score would have produced an adjusted score of 66?

3. *Critical Thinking* For some potential original scores, the adjusted score could actually be lower than the original score. Find an inequality, in terms of x, that represents the scores for which this will occur.

Alternative Assessment Rubric

For use after Chapter 2

JOURNAL
SOLUTION

1. Complete journal answers should include the following information:

   a. No *x*-value in the ordered pairs corresponds to 2 different *y*-values.

   b. Since no *x*-values correspond to 2 different *y*-values, then any vertical line should pass through the graph in *at most* one place.

   c. When substituting *x*-values into the equation, each *x*-value yields at most one *y*-value.

   d. Sample answers:

   - **a.** $(3, 4) (-3, 5)(0, -2) (1, 4)$
   - **b.** Any graph which passes the vertical line test.
   - **c.** $y = 2x^2 + 5x - 3$

MULTI-STEP
PROBLEM
SOLUTION

2. a. $(66, 76), (87, 90)$

b. $y = \dfrac{2}{3}x + 32$

c. 84

d. 51

3. $x > 96$

MULTI-STEP
PROBLEM
RUBRIC

4 Students complete all parts of the questions accurately. Student work and explanations show an understanding of writing linear functions and finding solutions of linear functions.

3 Students complete the questions and answers are mostly correct. Student is able to write and solve linear functions, but use of linear functions as an application is incorrect.

2 Students complete some questions correctly. Student is able to write or solve linear functions correctly.

1 Students' work is very incomplete. Solutions or reasoning are incorrect. Student work does not show an understanding of what a linear function is or how one is used.

Review and Assess

Project: Go, Car, Go!

For use with Chapter 2

OBJECTIVE **Determine how the height of a ramp affects the distance a toy car travels after going down the ramp.**

MATERIALS paper, pencil, graph paper, small toy car that rolls easily, books, ramp with sides (toy track or a piece of gutter)

INVESTIGATION *Collecting the Data* Stack the books and prop your ramp on them as shown in the diagram. Measure the height of the ramp at its highest point, to the nearest eighth of an inch. Set the car at the top of the ramp and let go. Do not push the car. Measure the horizontal distance the car traveled. Repeat two more times and then take the average of the distances found in the three trials. Record all data and computations in a table like the one below.

Ramp height	*Distance car traveled*			
	Trail 1	Trial 2	Trial 3	Average

Add another book, measure the new height of the ramp, collect and record data for three trials, and take the average. Continue until you have data for four or five ramp heights.

1. Draw a scatter plot of the height of the ramp and the average distance traveled data. Describe the correlation shown by the scatter plot and explain.

2. Approximate the best-fitting line for the data.

3. Choose a ramp height you have not used that is between the lowest and highest heights you did use. Use the fitted line to estimate how far the car would travel at this height. Test your prediction.

4. Choose a ramp height you have not used that is greater than the highest ramp height you did use. Use the fitted line to estimate how far the car would travel at this height. Test you prediction.

5. Which prediction was better?

PRESENT YOUR RESULTS Write a report about your experiment. Thoroughly describe the procedures you used. Include your data table, scatter plot, computations, equations, and predictions. Discuss the validity of your model for each type of prediction.

Review and Assess

Project: Teacher's Notes

For use with Chapter 2

GOALS
- Write an equation of a line.
- Use a scatter plot to identify the correlation shown by a set of data and approximate the best fitting line for a set of data.
- Use a mathematical model to make a prediction.

MANAGING THE PROJECT

You may wish to have students work in small groups to facilitate the data collection. Encourage groups to make collective decisions and to prepare the final report jointly. Make sure students use a ramp with sides to keep the car from falling off. Toy tracks or a piece of a gutter approximately 18 inches work well. You could also make a track out of poster board.

Encourage students to measure carefully and to the nearest eighth of an inch. You may wish to explain that the horizontal distance traveled by the car is the perpendicular distance from the bottom of the stack of books, below the edge of the ramp, to the back of the car. Important points to address are: measuring from the same reference points each time, setting the car in the same position each time before releasing it, and not pushing the car.

RUBRIC **The following rubric can be used to assess student work.**

4 The student's discussion of procedures indicates attention to detail and use of accurate and careful measurement techniques. The student plots the data correctly, chooses and finds an appropriate line of best fit, and makes appropriate predictions. The report presents an insightful comparison of the validity of the predictions and presents all procedures, data, and computations clearly and thoroughly.

3 The student describes procedures, makes the scatter plot, finds a line of best fit, and makes predictions. However, the student may not perform all calculations accurately, may not use careful measurement techniques, or may not fully address the issues when comparing the predictions. The report may not be as clear and thorough as possible.

2 The student describes procedures, makes the scatter plot, finds a line of best fit, and makes predictions. However, work may be incomplete or reflect misunderstanding. For example, the student may not measure accurately or to the correct precision or may not find an appropriate line of best fit. The report may indicate a limited grasp of certain ideas or may lack key information.

1 The discussion of procedures, scatter plot, line of best fit, or predictions are missing or do not show an understanding of key ideas. The report does not compare the predictions or fails to support the comparison made.

Review and Assess

NAME _____ DATE _____

Cumulative Review

For use after Chapters 1–2

Tell what property the statement illustrates. (1.1)

1. $4 + (-4) = 0$　　　　**2.** $2 + (5 + 7) = (2 + 5) + 7$　　**3.** $3(2 + 4) = 3(2) + 3(4)$

Select and perform an operation to answer the question. (1.1)

4. What is the sum of 13 and -6?　　　　　　**5.** What is the difference of 23 and -8?

6. What is the product of 4 and -8?　　　　　**7.** What is the quotient of 4 and $-\frac{2}{3}$?

Evaluate the expression for the given value of x. (1.2)

8. $x + 8$ when $x = -2$　　　　　　　**9.** $x^2 - 2$ when $x = -3$

10. $3x^2 - x + 1$ when $x = 5$　　　　　**11.** $3x^4 + x - 1$ when $x = 2$

Solve the equation. (1.3)

12. $2x + 1 = 8$　　　　　**13.** $2a + 1 = 4a - 8$　　　　**14.** $6x = 4 - 10x$

15. $4.5a - 1.7 = 7.3$　　　**16.** $3(2a - 8) = 8a - 12$　　　**17.** $\frac{1}{3}x + 4 = \frac{7}{9}x$

Solve the formula for the indicated variable. (1.4)

18. Area of trapezoid　　　　　　　**19.** Circumference of circle

Solve for b_1. $A = \frac{1}{2}(b_1 + b_2)$　　　　　Solve for r. $C = 2\pi r$

Solve the compound inequality. Graph its solution. (1.6)

20. $3x + 7 > 10$ or $-2x > 4$　　　　**21.** $15 < 5x < 25$

22. $-0.3 \le 0.2x + 0.5 \le 0.9$　　　　**23.** $-3x + 1 < -11$ or $5x + 2 < 7$

Solve the absolute value equation or inequality. (1.7)

24. $|3x - 4| = 12$　　　　**25.** $\left|\frac{1}{2}x - 6\right| = 8$　　　　**26.** $|7 - 2x| = 13$

27. $|x + 2| > 6$　　　　　**28.** $|6x - 4| \dagger 8$　　　　**29.** $|3 - 5x| < 13$

Evaluate the function when x = 2. (2.1)

30. $f(x) = x$　　　　　**31.** $g(x) = 5x$　　　　　**32.** $r(x) = x^2$

33. $g(x) = 3x - 5$　　　**34.** $h(x) = -2x^2 + 1$　　　**35.** $j(x) = x^3 + 2x^2$

Use the vertical line test to determine whether the relation is a function. (2.1)

36.
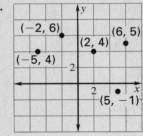

37.
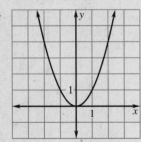

38.
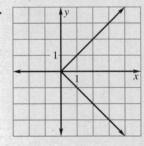

NAME _____ DATE _____

Cumulative Review

For use after Chapters 1–2

Tell whether the lines are *parallel, perpendicular,* or *neither.* (2.2)

39. Line 1: through $(-1, 8)$ and $(7, 9)$

Line 2: through $(2, 5)$ and $(10, 6)$

40. Line 1: through $(3, 4)$ and $(-5, 8)$

Line 2: through $(1, 2)$ and $(3, 6)$

Graph the equations. (2.3)

41. $y = 3x - 4$

42. $y = -\frac{2}{3}x + 5$

43. $3x - 2y = 4$

44. $x = 5$

45. $y = -\frac{2}{3}$

46. $5x - 10y = 20$

Write the equation of the line that passes through the given points. (2.4)

47. $(4, 2)$ and $(7, 8)$

48. $(5, -2)$ and $(-3, 4)$

49. $(4, 0)$ and $(-1, 8)$

50. $(-2, 1)$ and $(3, 1)$

51. $(4, 5)$ and $(4, 9)$

52. $(0, 5)$ and $(-5, 0)$

The variables *x* and *y* vary directly. Write an equation that relates the variables. Then find *y* when *x* = 5. (2.4)

53. $x = 3, y = 7$

54. $x = -2, y = 4$

55. $x = \frac{1}{2}, y = 4$

56. $x = 8, y = -2$

57. $x = -6, y = -6$

58. $x = 0.2, y = 0.8$

Draw a scatter plot of the data. Then approximate the best-fitting line for the data. (2.5)

59.

x	−3	−3	0	1	1	2	4
y	−7	−3	1	−1	5	1	5

Graph the inequality in a coordinate plane. (2.6)

60. $5x - 2y > 10$

61. $4x < -20$

62. $8y > 10$

63. $y > \frac{1}{3}x + 3$

64. $0.25x + 1 > 2$

65. $3x < -\frac{1}{2}y$

Evaluate the function for the given value of *x*. (2.7)

$f(x) = \begin{cases} 3x, & \text{if } x > 5 \\ -x + 2, & \text{if } x \leq 5 \end{cases}$

66. $f(3)$

67. $f(8)$

68. $f(-3)$

69. $f(5)$

70. $f(0)$

71. $f\left(\frac{19}{3}\right)$

Graph the function. Then identify the vertex, tell whether the graph opens up or down, and tell whether the graph is wider, narrower, or the same width as the graph of $y = |x|$. (2.8)

72. $y = |x| + 7$

73. $y = -|x| + 8$

74. $y = -2|x + 2| + 1$

75. $y = |x| - \frac{3}{2}$

76. $y = \frac{1}{2}|x| + 2$

77. $y = |x| - 4$

Review and Assess

ANSWERS

Chapter Support

Parent Guide

2.1: No; for the input 0 there are two outputs, 1 and -1. **2.2:** 2.5 inches per year
2.3: $8x + 10y = 240$; x-intercept: 30, y-intercept: 24 **2.4:** $y = \frac{1}{3}x - 2$ **2.5:** negative correlation **2.6:** yes **2.7:** three T-shirts: $24, six T-shirts: $30 **2.8:** $(5, 2)$; down

Prerequisite Skills Review

1. 3 **2.** -3 **3.** $-\frac{3}{2}$ **4.** -6 **5.** $y = \frac{1}{2}x - 4$
6. $y = 2x - 5$ **7.** $y = -\frac{2}{7}x + 7$
8. $y = -2x - 15$ **9.** $x > -3$ **10.** $y \le -1$
11. $x \le 3$ **12.** $y > 10$

Strategies for Reading Mathematics

1. domain: $-4, -2, 0, 2, 4$;
range: $0, 2, 4, 6, 8$; -4

2.

input	-4	-4	-2	0	2	2	4
output	0	2	2	4	4	6	8

3. domain: $-3, -1, 1, 2, 4, 5$; range: $-5, -1, 1, 2, 3, 4$; not a function; 1 maps to both 1 and 2

4. 0, 1, 4, 9; *Sample answer:*

input	-3	-2	-1	0	1	2	3
output	9	4	1	0	1	4	9

The relation is a function.

Lesson 2.1

Warm-Up Exercises

1. -10 **2.** 1 **3.** 14 **4.** 8 **5.** 1

Daily Homework Quiz

1. $2 - 5x = -6$ or $2 - 5x = 6$ **2.** no
3. $\frac{1}{3}, 1$ **4.** $-7 < 5 + 4x < 7$
5. $1 < x < \frac{7}{2}$

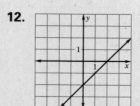

6. $x \le -7$ or $x \ge -3$

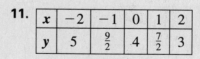

Lesson Opener

Allow 5 minutes.

B, C, D, and F are functions.

Practice A

1. domain: $\{-1, 0, 2\}$; range: $\{3, 6, 16\}$
2. domain: $\{3, 4, 9\}$; range: $\{-9, 0\}$
3. domain: $\{1, 2\}$; range: $\{-12, 6, 24\}$

4.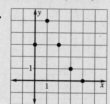

5.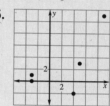

The relation is a function.

The relation is *not* a function.

6.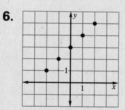

The relation is a function.

7. The relation is a function. **8.** The relation is *not* a function. **9.** The relation is a function.

10.

x	-2	-1	0	1	2
y	-1	1	3	5	7

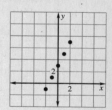

11.

x	-2	-1	0	1	2
y	5	$\frac{9}{2}$	4	$\frac{7}{2}$	3

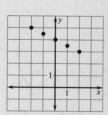

12.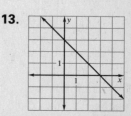

13.

Lesson 2.1 *continued*

14.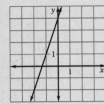

15.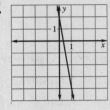

3. The relation is a function.   **4.** The relation is *not* a function.   **5.** The relation is a function.

6.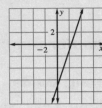

7.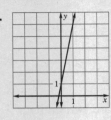

16.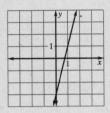

17.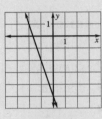

8.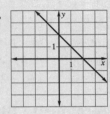

9.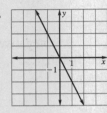

18.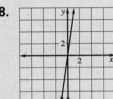

19.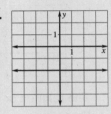

10.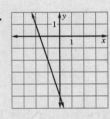

11.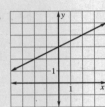

20.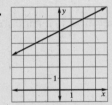

12.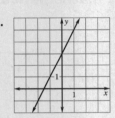

13.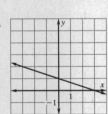

21.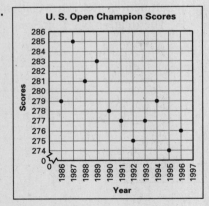

14.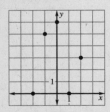

15. linear; 4
16. not linear; 2
17. linear; -2
18. not linear; 14
19. not linear; $\frac{1}{2}$
20. linear; 5

21. 54; $S(3)$ represents the surface area of a cube with sides of length 3.   **22.** $\{6, 7, 9, 10\}$

23. $\{6, 7, 9, 10\}$ **24.** yes

Practice B

1.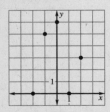

2.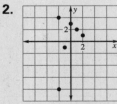

The relation is a function.

The relation is *not* a function.

Practice C

1. The relation is *not* a function.   **2.** The relation is a function.   **3.** The relation is *not* a function.   **4.** First quadrant   **5.** Second quadrant   **6.** Third quadrant   **7.** Fourth quadrant

Lesson 2.1 *continued*

8.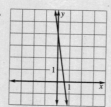

9.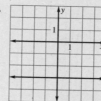

10.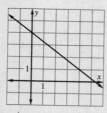

11.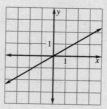

12.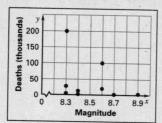

13.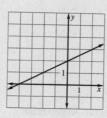

14. linear; 16 **15.** not linear; -1

16. not linear; 0 **17.** not linear; 49

18. not linear; $-\frac{5}{6}$ **19.** not linear; -2

20. Domain $= \{8.3, 8.4, 8.6, 8.7, 8.9\}$

Range $= \{1530, 2000, 2990, 5000, 10,700,$
$20,000, 28,000, 100,000, 200,000\}$

21.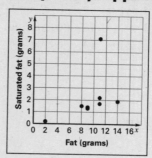

22. No. For each input there is not exactly one output. For example 200,000, 28,000, and 5000 are all outputs for the input 8.3.

Reteaching with Practice

1. $D = \{-3, -1, 1, 2\}$
   $R = \{0, 2, 5\}$
   function

2. $D = \{-3, -2, 5\}$
   $R = \{6, 7, 8\}$
   function

3. $D = \{-4, -2, 0\}$
   $R = \{-2, -1, 3, 4\}$
   not a function

4.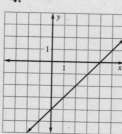

5.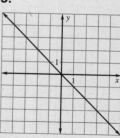

6.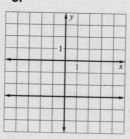

7.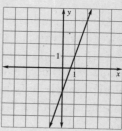

8.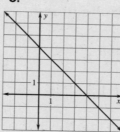

9.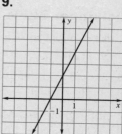

10. -11 **13.** -3 **14.** 11

Interdisciplinary Application

1. a.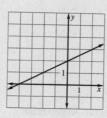

b. No, this graph is not the graph of a function. Although answers will vary, an example reason could be that there are several points corresponding to the single *x*-value 11.

c. Answers may vary, but the points for Butter and French Dressing (Low Calorie) appear to be separated from the rest of the foods.

Lesson 2.1 *continued*

d. Olive Oil, Blue Cheese Dressing, French Dressing (Regular and Low Calorie), Italian Dressing, and Mayonnaise.

2. a.

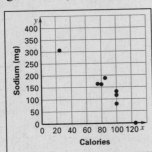

b. No, this graph is not a function. Although answers will vary, an example reason could be that there are several points corresponding to the single *x*-value 100.   **c.** Answers may vary, but Low Calorie French Dressing seems to be separated from the rest of the foods.   **d.** Olive oil and mayonnaise.

Challenge: Skills and Applications

1. no; *Sample answer:*
$f(1) + f(2) = 1 + 4 = 5 \neq f(1 + 2) = 9$

2. yes   **3.** no; *Sample answer:*
$f(1) + f(-1) = 1 + 1 = 2 \neq f(1 + (-1)) = 0$

4. no; *Sample answer:*
$f(1) + f(2) = 6 + 11 = 17 \neq f(1 + 2) = 16$

5. $b = 0$ **6.** even   **7.** neither   **8.** odd

9. even   **10.** The coefficients of the odd-power terms must be 0; the coefficients of the even-power terms must be 0.   **11.** yes   **12.** no; *Sample answer:* $f(1) = f(-1) = 1$ **13.** no; *Sample answer:* $f(2) = f(-2) = 2$ **14.** yes

Lesson 2.2

Warm-Up Exercises

1. $-\frac{5}{7}$ **2.** 2   **3.** $-\frac{1}{3}$ **4.** $-\frac{4}{3}$

Daily Homework Quiz

1. D: $-2, -1, 0, 1, 2$; R: $-2, -1, 1, 2$

2. 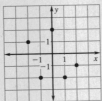   ; function

3. 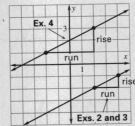   **4.** not linear; 34

Lesson Opener

Allow 10 minutes.

1. $\frac{2}{3}$ **2.**

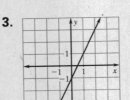

3. See graph above; $\frac{1}{2}$ **4.** See graph above; $\frac{1}{2}$
5. lines are parallel; slopes are the same

Graphing Calculator Activity

1. $y = 2x$ and $y = 7x$

2. $y = -5x$ and $y = -0.3x$

4. $y = 7x, y = 5x, y = 3x, y = 0.7x, y = 0.3x$

Practice A

1. 2   **2.** 0   **3.** -1 **4.** 2   **5.** -3 **6.** 8

7. $-\frac{1}{2}$ **8.** $-\frac{1}{3}$ **9.** $\frac{4}{3}$ **10.** rises

11. is horizontal   **12.** falls   **13.** rises   **14.** falls

15. is vertical   **16.** neither   **17.** neither

18. perpendicular   **19.** parallel

20. perpendicular   **21.** neither

22. 12.5 quarts per hour   **23.** Yes, $\frac{3}{200} < \frac{1}{64}$

Practice B

1. -2 **2.** -1 **3.** $-\frac{1}{3}$ **4.** $\frac{3}{2}$ **5.** -1 **6.** $\frac{3}{2}$

7. Line 1: $m = 1$; Line 2: $m = 4$; Line 2 is steeper than Line 1.

Lesson 2.2 *continued*

8. Line 1: $m = -2$; Line 2: $m = -3$; Line 2 is steeper than Line 1.

9. Line 1: $m = -\frac{1}{2}$; Line 2: $m = -1$; Line 2 is steeper than Line 1.

10. Line 1: $m = \frac{1}{3}$; Line 2: $m = \frac{1}{8}$; Line 1 is steeper than Line 2.

11. $m = 1$; rises   **12.** $m = 0$; is horizontal

13. $m = -\frac{1}{4}$; falls   **14.** m is undefined; is vertical   **15.** $m = -\frac{5}{3}$; falls   **16.** $m = 0$; is horizontal   **17.** perpendicular

18. perpendicular   **19.** neither

20. perpendicular   **21.** $\frac{1}{12}$

22. 0.85 ticket per minute

Practice C

1. $\frac{2}{5}$ **2.** $\frac{3}{2}$ **3.** $-\frac{17}{11}$ **4.** $-\frac{1}{4}$ **5.** -5 **6.** $-\frac{16}{17}$

7. rises   **8.** is vertical   **9.** falls   **10.** Line 2

11. Line 1   **12.** Line 2   **13.** Line 1

14. They are equal.   **15.** They are negative reciprocals of each other.   **16.** vertical lines

17. horizontal lines   **18.** $\frac{1000}{21}$; $\frac{55}{17}$ **19.** $\frac{964}{755}$; $\frac{180}{151}$

20.

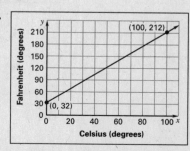

This is true for an equilateral triangle of any size.

Reteaching with Practice

1. 1; rises   **2.** $-\frac{1}{5}$; falls   **3.** 0; horizontal

4. undefined; vertical   **5.** undefined; vertical

6. $\frac{3}{2}$; rises   **7.** neither   **8.** perpendicular

9. 1.5°F per hour decrease; 82.5°F

Cooperative Learning Activity

Instructions

1–2.

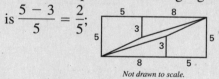

---

3. $\frac{9}{5}$ **4.** $(0, 32)$ **5.** $F = \frac{9}{5}C + 32$

Analyzing

1. Yes–by locating points on the graph or by substituting values of C in the equation.

2. Yes–since you have an ordered pair, you have both Fahrenheit and Celsius values.

Real–Life Application

1. 0.07635, 0.01895, 0.08, 0.0246   **2.** Wages vs. Furniture Sales   **3.** Wages vs. Auto Sales

4. a. Furniture Sales   **b.** Auto Sales   **5.** Yes; this says that an increase in any sector of the economy is correlated with increases in national wages.   **6.** This would mean that an increase in sales in some sector of the economy is correlated with a decrease in national wages.

7. This would mean that the sector does not help in predicting national wages.

Challenge: Skills and Applications

1. yes   **2.** no   **3.** $\frac{6}{7}$ **4.** 2

5. a. area of square = 64; area of rectangle = 65 **b.** Each figure seems to be made up of 4 non-overlapping polygons that are congruent to the 4 nonoverlapping polygons of the other.   **c.** The slope of the hypotenuse of the right triangle is $\frac{3}{8}$, while the slope of the slanting leg of the trapezoid is $\frac{5 - 3}{5} = \frac{2}{5}$;

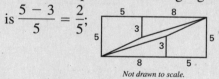

Not drawn to scale.

6. a. As a approaches 0, the slope of the line s gets closer and closer to the slope of line t.

b. $(1 + a)^2 = 1 + 2a + a^2$

c. $\dfrac{(1 + 2a + a^2) - 1}{(1 + a) - 1} = \dfrac{2a + a^2}{a} = 2 + a$

d. The slope is 2.

Lesson 2.3

Warm-Up Exercises

1. 6 **2.** 8 **3.** 6 **4.** $y = -2x + 150$

5. $y = \frac{8}{3}x - 2$

Lesson 2.3 *continued*

Daily Homework Quiz

1. -9; falls **2.** 0; is horizontal **3.** Line 1
4. Line 2 **5.** perpendicular **6.** 2 ft/sec

Lesson Opener

Allow 15 minutes.

1. 4 **2.** $2x + 3y = 30$

3.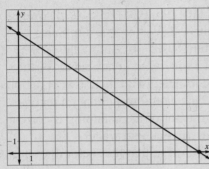

4. $(15, 0)$; Lisa made 15 two-point field goals if she didn't make any three point ones.

5. $(0, 10)$; Lisa made 10 three-point field goals if she didn't make any two-point ones.

Practice A

1. 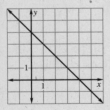 **2.**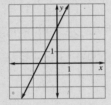

3. 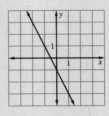 **4.**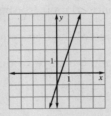

5.  **6.**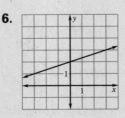

7. 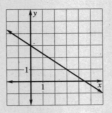 **8.**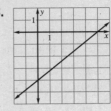

9. 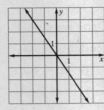 **10.** $m = 3$; $b = 1$

11. $m = -4$; $b = 7$ **12.** $m = 6$; $b = -4$
13. $m = -8$; $b = -2$ **14.** $m = \frac{5}{3}$; $b = 1$
15. $m = -\frac{1}{5}$; $b = -3$

16. 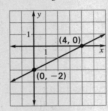 **17.**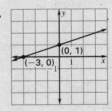

18. 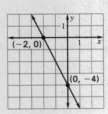 **19.**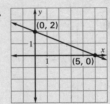

20. 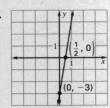 **21.**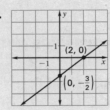

22. 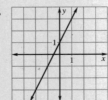 **23.**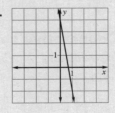

Lesson 2.3 *continued*

24. 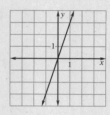   **25.**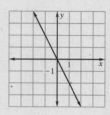

26. 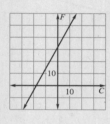   **27.**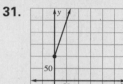

28. 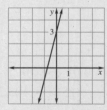   **29.** 3   **30.** 100

31.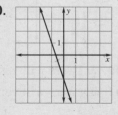

Practice B

1. $m = 8; \ b = -7$ **2.** $m = -10; \ b = 0$

3. $m = -\frac{1}{4}; \ b = \frac{3}{2}$ **4.** $m = -\frac{1}{2}; \ b = \frac{1}{4}$

5. $m = \frac{3}{7}; \ b = \frac{5}{7}$ **6.** $m = \frac{2}{3}; \ b = 2$

7. x-intercept: $\frac{1}{3}$; y-intercept: -1

8. x-intercept: 6; y-intercept: 6

9. x-intercept: -3; y-intercept: 2

10. x-intercept: 12; y-intercept: 3

11. x-intercept: $\frac{12}{5}$; y-intercept: -4

12. x-intercept: $-\frac{6}{7}$; y-intercept: -3

13. x-intercept: 2; y-intercept: -4

14. x-intercept: -4; y-intercept: 3

15. x-intercept: $-\frac{8}{5}$; y-intercept: -4

16. x-intercept: 4; y-intercept: $-\frac{4}{3}$

17. x-intercept: -4; y-intercept: $-\frac{8}{5}$

18. x-intercept: $-\frac{1}{2}$; y-intercept: 3

19.    **20.**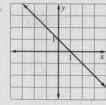

21. 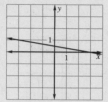   **22.**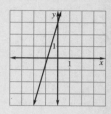

23. 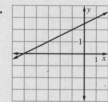   **24.**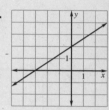

25. 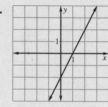   **26.**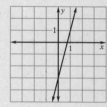

27. 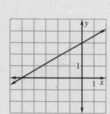   **28.** $\frac{2}{7}$ **29.** 2

30. $y = \frac{2}{7}x + 2$ **31.** $0.10d + 0.25q = 50$

32. $0.03x + 0.04y = 250$

Practice C

1. $m = 4; \ b = 2$ **2.** $m = -3; \ b = \frac{1}{2}$

3. $m = -\frac{2}{3}; \ b = 4$ **4.** $m = 2; \ b = -3$

5. $m = 0; \ b = 6$ **6.** $m = \frac{4}{3}; \ b = \frac{1}{3}$

7. $m = \frac{-7}{5}; \ b = \frac{8}{5}$ **8.** $m = \frac{3}{2}; \ b = -2$

9. $m = \frac{8}{3}; \ b = 0$ **10.** $m = \frac{2}{5}; \ b = \frac{7}{5}$

11. $m = \frac{3}{7}; \ b = \frac{1}{7}$ **12.** $m = -\frac{1}{2}; \ b = \frac{5}{2}$

13. x-intercept: 4   **14.** x-intercept: -4
     y-intercept: 3 y-intercept: 8

15. x-intercept: $\frac{5}{3}$ **16.** x-intercept: 0
     y-intercept: $\frac{5}{2}$ y-intercept: 0

17. x-intercept: $\frac{3}{4}$ **18.** x-intercept: -13
     y-intercept: 3 y-intercept: none

19. x-intercept: none   **20.** x-intercept: -1
     y-intercept: $\frac{3}{4}$ y-intercept: $\frac{1}{4}$

21. x-intercept: $-\frac{1}{2}$
     y-intercept: $-\frac{1}{4}$

Lesson 2.3 *continued*

22.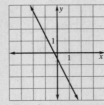

23.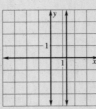

24.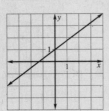

25.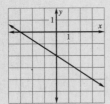

26.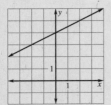

27.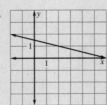

28.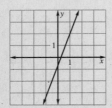

29.

30.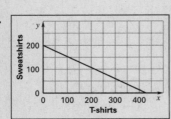

31.

$7x + 15y = 3000$

Sample answers:
(210, 102)
(300, 60)
(390, 18)

32. a. 300,000; The *V*-intercept represents the initial value of the equipment.

b. $\frac{-100,001}{2} = -50,000.5$; The slope represents the decrease in value per year.

Reteaching with Practice

1.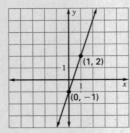

2.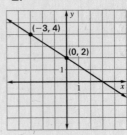

3.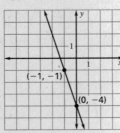

4.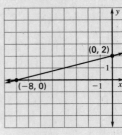

5.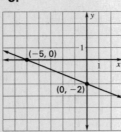

6.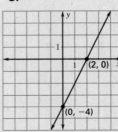

7.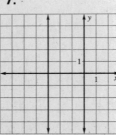

8.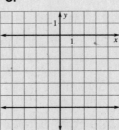

9.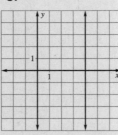

10. answers will vary; $1.25h + 2.5p = 650$

11. answers will vary; $5.35r_1 + 6.25r_2 = 105.3$

Lesson 2.3 *continued*

Interdisciplinary Application

1. 0.003 **2.** *Sample Answer:* Each word in the document increases the load time of the document by 0.03 second. **3.** 2.2 **4.** *Sample Answer:* If there were no words in the document, this model predicts that the document would still take 2.2 seconds to load. This could be because of background graphics, or because the HTML programmers were using only approximate numbers in their calculations of the equation.

5. $-0.003x + y = 2.2$

6.

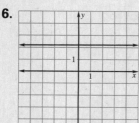

7. 320 sec **8.** 47 sec
9. 71 sec

10. *Flatland* **11.** *Sample Answer:* Longer works could be split into several different documents, each loading in less than the required 60 seconds.

Math and History Applications

1. About 51 hours after the near miss, on the morning of April 17 **2.** About 2.45 times as fast **3.** About 2 knots **4.** The graph won't be linear: it will consist of upward segments (swimming) followed by more or less horizontal ones (drifting), and it might even have negative slope if currents or wind carried the swimmer or the boat back toward the North American coast.

Challenge: Skills and Applications

1. 13 **2.** 3 **3.** $y = -\dfrac{q}{p}x + q$ **4.** The given equation implies that $x_1y_2 = x_2y_1$. Therefore, $x_1(mx_2 + b) = x_2(mx_1 + b)$, or $x_1x_2m + x_1b = x_1x_2m + x_2b$; $(x_1 - x_2)b = 0$. Since $x_1 \ne x_2$, $b = 0$.

5. a. $y = \dfrac{r}{p}x + s - \dfrac{qr}{p}$; $m = \dfrac{r}{p}$, $b = s - \dfrac{qr}{p}$

b. $t = \dfrac{q - x}{p}$; $y = s - r\left(\dfrac{q - x}{p}\right) = \dfrac{r}{p}x + s - \dfrac{qr}{p}$; the particle is at different locations along the same line at time t. **6.** $y = mx + (b - 3m)$; the new intercept is $b - 3m$ **7.** $y = mx + (b + 2m + 5)$

Quiz 1

1. Domain: $\{-5, -4, 2, 5\}$
  Range: $\{-6, 1, 3, 4, 5\}$
  function: no

2. no **3.** 38 **4.** parallel **5.** perpendicular

6.

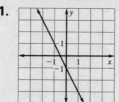

7.

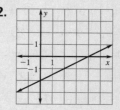

Lesson 2.4

Warm-Up Exercises

1. $2; -4$ **2.** $\dfrac{3}{2}; 1$ **3.** 2 **4.** $-\dfrac{2}{3}$

Daily Homework Quiz

1.

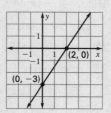

2.

3. $3; -8$ **4.**

Lesson Opener

Allow 10 minutes.

1. 0.06, 0.06, 0.06, 0.06 **2.** $V = 0.06p$
3. 10, 5, 7.5, 3 **4.** no; *Sample explanation:* The value of k would need to be different for the different activities.

Practice A

1. $y = 3x + 2$ 2. $y = 4x - 5$
3. $y = -6x + 1$ 4. $y = -x - 9$
5. $y = 2x$ 6. $y = 7$ 7. $y = 5x + 3$
8. $y = 3x - 2$ 9. $y = -2x + 4$
10. $y = 4x - 11$ 11. $y = 6$ 12. $y = x - 5$
13. $y = -x + 3$ 14. $y = 2x - 1$
15. $y = -x + 10$ 16. $y = 2x - 1$
17. $y = -8x + 17$ 18. $y = 4x + 8$
19. $y = x + 1$ 20. $y = -3x + 20$
21. $y = -\frac{3}{4}x + 3$ 22. $y = 5x - 11$
23. $y = -2x + 22$ 24. $y = -\frac{1}{2}x + \frac{17}{2}$
25. $y = -3x + 15$ 26. $y = 2x + 3$
27. $y = -\frac{1}{3}x + \frac{1}{3}$
28. The data do not show direct variation.
29. The data do show direct variation, and the direct variation equation is $y = -x$.
30. $y = 0.06x$

Practice B

1. $y = 4x - 4$ 2. $y = -6x + 3$
3. $y = \frac{4}{3}x + 6$ 4. $y = -\frac{1}{2}x - 4$ 5. $y = 8x$
6. $y = 5$ 7. $y = -2x + 5$ 8. $y = 5x + 23$
9. $y = x - 12$ 10. $y = 3x - 7$
11. $y = -8x + 8$ 12. $y = -4x + \frac{8}{3}$
13. $y = \frac{3}{4}x + \frac{5}{2}$ 14. $y = -2x + 1$
15. $y = \frac{1}{2}x + \frac{1}{2}$ 16. $y = -\frac{1}{2}x$ 17. $y = x + 4$
18. $y = -2x - 1$ 19. $y = 3x + 19$
20. $y = \frac{3}{7}x - \frac{5}{7}$ 21. $y = \frac{3}{2}x - 19$
22. $y = 3x;\ 30$ 23. $y = -5x;\ -50$
24. $y = -\frac{5}{2}x;\ -25$ 25. $y = 0.25x;\ 2.5$
26. $y = -\frac{1}{4}x;\ -\frac{5}{2}$ 27. $y = \frac{27}{10}x;\ 27$
28. $y = \frac{103}{64}x$ 29. Yes, you are traveling about 88.5 km/hr.   30. $y = 0.2t + 14.7$
31. 16.7 pounds

Practice C

1. $y = 9x - 19$ 2. $y = -\frac{1}{7}x + \frac{16}{7}$ 3. $x = 1$

4. $y = \frac{8}{11}x - \frac{34}{11}$ 5. $y = 8$ 6. $y = x$
7. $y = -\frac{1}{2}x + \frac{7}{2}$ 8. $y = \frac{1}{4}x + \frac{11}{4}$

9. $y = -2x + 3$ 10. $y = \frac{3}{2}x + \frac{11}{2}$
11. $x = 7$ 12. $y = 2$ 13. $y = 2x + 5$
14. $y = -x$ 15. $y = x - 2$ 16. $y = \frac{1}{2}x - \frac{11}{2}$
17. $y = -\frac{3}{4}x - \frac{9}{2}$ 18. $y = -9$
19. $y = 0.49t + 31.4$
20. Yes. The model gives 41.2%, and the actual number was 41.1%.   21. No. The model gives 46.1%, and the actual number was 47%.
22. $y = 1420t + 15,500$
23. $y = 1420t - 2,810,300$ 24. $29,700; yes.

Reteaching with Practice

1. $y = 3x$ 2. $y = \frac{3}{4}x + 2$ 3. $y = -2x - 3$
4. $y = -5x + 9$ 5. $y = \frac{1}{3}x + 5$
6. $y = -2$ 7. $y = -3x + 11$
8. $y = x + 3$ 9. $y = -\frac{1}{5}x - 1$
10. $y = 10x;\ -20$ 11. $y = -4x;\ 8$
12. $y = -\frac{1}{9}x;\ \frac{2}{9}$

Real–Life Application

1. $(0, 0), (22, 100)$ 2. $y = 4.55x$
3. Wade 68.3; Felecia 81.9; Lee 45.5; Brandy 36.4; Tonya 95.6   4. *Sample answer:* Students may feel that a scale of 0 to 100 is easier to interpret because of its similarity to grades and percentages.   5. $(0, 10), (22, 1)$
6. $y = -0.409x + 10$ 7. Wade 3.9; Felecia 2.6; Lee 5.9; Brandy 6.7; Tonya 1.4

Challenge: Skills and Applications

1. $-\frac{27}{2}$ 2. -3 3. $\dfrac{x_2y_1 - x_2y_2}{x_2 - x_1}$

4. a. $(3, -2), (-5, 2);\ y + 2 = -\frac{1}{2}(x - 3)$ or

$y - 2 = -\frac{1}{2}(x + 5)$ b. $(-1, 0)$; it is the midpoint of the segment joining the two points; *Sample answer:* any such point is on the segment joining the two points.
5. $x_2(y_1 - b) = x_2(mx_1 + b - b) = mx_1x_2$; on the other hand, $x_1(y_2 - b) = x_1(mx_2 + b - b) = mx_1x_2$ 6. all lines passing through the point $(0, b)$ 7. all lines passing through the point (x_1, y_1); yes

Lesson 2.5

Lesson 2.5

Warm-Up Exercises

1. 1 **2.** $y = -2x + 1$ **3.** $y = 10x - 500$

4. $y = 13x + 5.12$

Daily Homework Quiz

1. $y = 4x - 2.3$ **2.** $y = -x - 1$

3. $y = \frac{1}{3}x - 6$ **4.** $y = -2x + 1$

5. $y = 2.5x$; 7.5 **6.** $y = \frac{1}{12}x$; $\frac{1}{4}$

Lesson Opener

Allow 5 minutes.

1. yes; increase **2.** positive correlation

3. negative correlation **4.** positive correlation

5. positive correlation **6.** negative correlation

7. negative correlation

Practice A

1. positive correlation **2.** negative correlation

3. relatively no correlation

4. 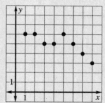 **5.**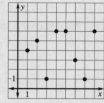

negative correlation   relatively no correlation

6. 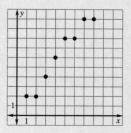 positive correlation

7. Answers may vary. *Sample*: $y = -\frac{3}{4}x + \frac{7}{2}$

8. Answers may vary. *Sample*: $y = \frac{1}{4}x + \frac{3}{2}$

9.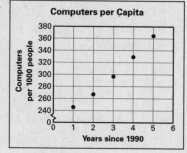

positive correlation

Practice B

1. 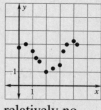 **2.**

positive correlation   relatively no correlation

negative correlation

3.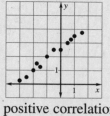

4. Answers may vary. *Sample*: $y = -\frac{2}{5}x + \frac{11}{4}$

5. Answers may vary. *Sample*: $y = x + \frac{1}{4}$

6. 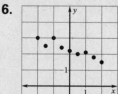 Answers may vary. *Sample*: $y = -\frac{1}{3}x + \frac{7}{3}$

7. 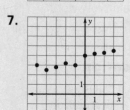 Answers may vary. *Sample*: $y = \frac{1}{4}x + \frac{15}{4}$

8.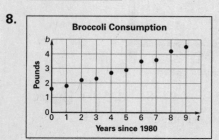

Lesson 2.5 *continued*

9. *Sample answer*: $b = 0.3t + 1.5$

10. *Sample answer*: 8.1 pounds

Practice C

1. $y = 7.6x + 4.5$ **2.** $y = -1.2x - 3.5$

3. Answers may vary. **4.** Answers may vary.

Sample: $y = \frac{-3}{5}x + \frac{31}{5}$ *Sample:* $y = x + 2.5$

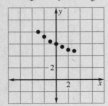 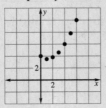

5. Answers may vary. **6.**

Sample: $3.3x + 0.87$

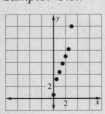 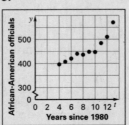

7. Answers may vary. **8.** Answers may vary.
Sample: *Sample*:
$y = 16.05t + 319.44$ approximately 640

9. **10.** Answers may
vary. *Sample:*
$y = 0.43x + 45$

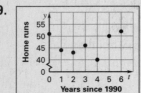

Reteaching with Practice

1. positive correlation **2.** relatively no
correlation **3.** $y = -2x + \frac{15}{2}$ **4.** $y = \frac{8}{3}x + 24$

Interdisciplinary Application

1. 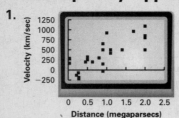 ; positive
correlation

2. $y = 454.158x - 40.784$ **3.** 549.6 km/sec

4. about 0.64 megaparsecs

Challenge: Skills and Applications

1. a. $(32, 43)$ **b.** $(0,0)$; if the mean is subtracted
from each point, the mean of the new data is 0.

2. a. -1.46 **b.** $y - 43 = -1.46(x - 32)$

3. a. 8.124, 11.979 **b.** the standard deviations
are the same; the standard deviation of translated
data is the same as that of the original data.

Quiz 2

1. $y = \frac{3}{4}x + 4$ **2.** $y = 4x + 3$

3. $y = -\frac{1}{3}x + \frac{1}{3}$ **4.** $y = -x - 1$

5. positive correlation **6.** relatively no
correlation **7.** negative correlation

Lesson 2.6

Warm-Up Exercises

1–4.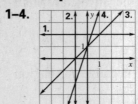

Daily Homework Quiz

1. 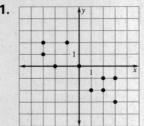 negative correlation

2. 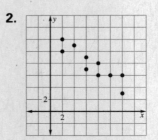 *Sample answer:*
$y = -\frac{2}{3}x + 12\frac{2}{3}$, using
$(4, 10)$ and $(10, 6)$

3. *Sample answer*: about $3\frac{1}{3}$

Lesson Opener

Allow 20 minutes.

Check students' graphs.

Lesson 2.6 *continued*

Answers

Practice A

1. yes; no **2.** no; yes **3.** no; yes

4. yes; yes **5.** yes; no **6.** yes; no

7. 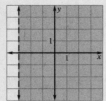 **8.**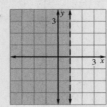

9. 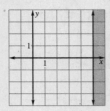 **10.**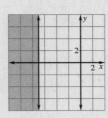

11. 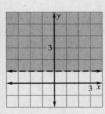 **12.**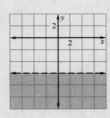

13. 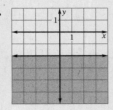 **14.**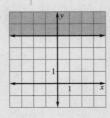

15. D **16.** F **17.** C **18.** E **19.** B **20.** A

21. $2x + 3y \geq 34$ **22.** no **23.** Answers may vary. *Sample:* 13 2-point and 3 3-point or 17 2-point and 0 3-point.

Practice B

1. no; yes **2.** no; yes **3.** no; yes

4. yes; no **5.** yes; no **6.** no; yes

7. 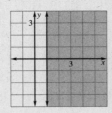 **8.**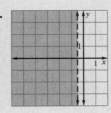

9. 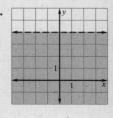 **10.**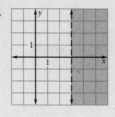

11. 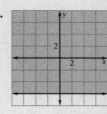 **12.**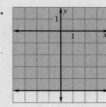

13. 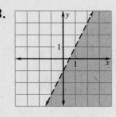 **14.**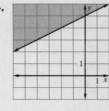

15. 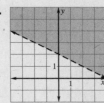 **16.**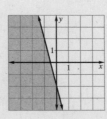

17. 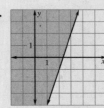 **18.**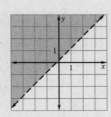

19. 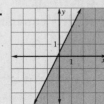 **20.**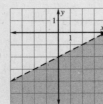

21.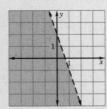

Copyright © McDougal Littell Inc.
All rights reserved.

Algebra 2
Chapter 2 Resource Book

A13

Lesson 2.6 *continued*

22. $t \le 5p$

23. $(2, 12)$

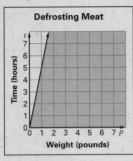

Defrosting Meat

24. Yes, a 2-pound roast takes at most 10 hours to defrost, so it will be completely defrosted before 12 hours passed.

25. $3x + 5y \le 800$

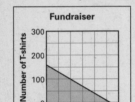

Fundraiser

26. $(50, 150)$

27. No, $(50, 150)$ is not a solution of $3x + 5y \le 800$.

Practice C

1.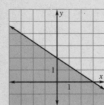

2.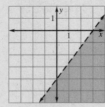

3.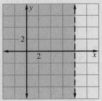

4.

5.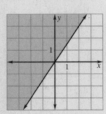

6.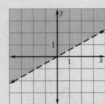

7.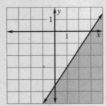

8.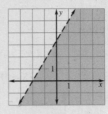

9.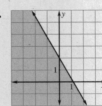

10.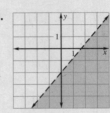

11.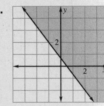

12.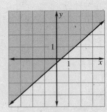

13.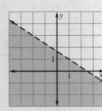

14.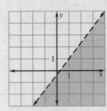

15.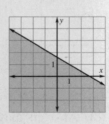

16. $4x + 2y \le 92$

17.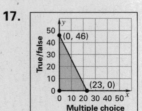

Multiple choice / True/false, (0, 46), (23, 0)

18. No.

19. $3T + 3.50S \ge 47.50$

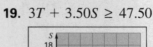

Thompsons / Stewarts

Lesson 2.6 continued

20. *Sample answers:* 10 hours at Thompson's and 10 hours at Stewart's, 15 hours at Thompson's and 5 hours at Stewart's, 5 hours at Thompson's and 10 hours at Stewart's

21. $y < -\frac{3}{5}x + 2$ **22.** $y \geq \frac{2}{3}x + \frac{1}{2}$

Reteaching with Practice

1. yes; yes   **2.** yes; no   **3.** yes; no

4. no; no

5.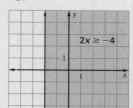

6.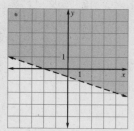

7.

8.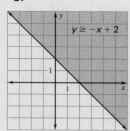

9.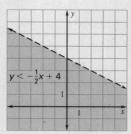

10.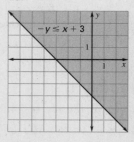

Real–Life Application

1. 120   **2.** $x + y \leq 120$ **3.** $x \leq 2y$

4. Food gathered $= 6x + 4y$

5.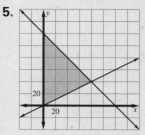

6. Answers will vary.
Sample answer:
(15, 30), (20, 40),
(10, 30), (25, 20);
(20, 40)

7. *Sample answer:* The optimum amount is at (80, 40)

Challenge: Skills and Applications

1. a. $x = x_0;\ \frac{1}{2}x_0 + 3$ **b.** $y > \frac{1}{2}x_0 + 3$

c. Solve the inequality for y; shade the region above the graph if the inequality sign is ">" or "≥"; otherwise, shade below the graph.

2.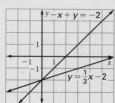

3.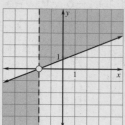

4. a. All lines through the point $(2, 5)$
b. $m < \frac{2}{5}$ **c.** $m < 0$ or $m > \frac{2}{5}$; $m < 0$
5. a. $y - b = \frac{3}{2}(x - a)$ **b.** $y \geq \frac{3}{2}(x - 4) + 5$

Lesson 2.7

Warm-Up Exercises

1. -8 **2.** -5 **3.**

Daily Homework Quiz

1.

2.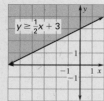

3.

Lesson 2.7 *continued*

Lesson Opener
Allow 10 minutes.

1. $1000 **2.** $1800 **3.** $1500 **4.** $1 \le x < 2$
5. $0 \le x \le 4$ **6.** $0, 400, 1000, 1500, 1800$

Practice A

1. 2 **2.** 3 **3.** 3 **4.** 2 **5.** 13 **6.** 5 **7.** 4
8. 8 **9.** -6 **10.** -5 **11.** 5 **12.** -9
13. E **14.** B **15.** D **16.** F **17.** C **18.** A

19. $f(x) = \begin{cases} 0, & 0 \le x \le 5 \\ 5, & 5 < x \le 12 \\ 12, & 12 < x \le 18 \\ 18, & x > 18 \end{cases}$

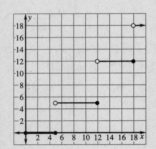

Practice B

1. -7 **2.** -1 **3.** -2 **4.** -16 **5.** -2
6. -7 **7.** 14 **8.** -7 **9.** -21 **10.** -9
11. 5 **12.** $\frac{5}{3}$

13. 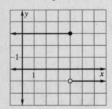 **14.**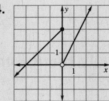

15. 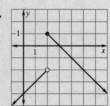 **16.**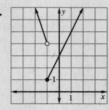

17. 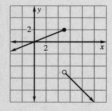 **18.**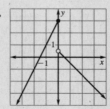

19. 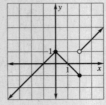 **20.**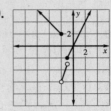

21.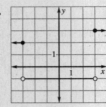

22. $f(x) = \begin{cases} -2x + 2, & \text{if } x \le 1 \\ x - 3, & \text{if } x > 1 \end{cases}$

23. $f(x) = \begin{cases} -2x + 2, & \text{if } x < 1 \\ x - 3, & \text{if } x \ge 1 \end{cases}$

24. $f(x) = \begin{cases} -2x + 2, & \text{if } x < 1 \\ -1, & \text{if } x = 1 \\ x - 3, & \text{if } x > 1 \end{cases}$

25. 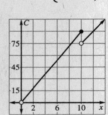 **26.** $x > 0; C > 0$

27. $C = \begin{cases} 0.05x, & \text{if } 0 < x \le 100{,}000 \\ 0.08x, & \text{if } x > 100{,}000 \end{cases}$

Practice C

1. 7 **2.** 2 **3.** 6 **4.** 6 **5.** 3 **6.** 1 **7.** -3
8. -7 **9.** 8 **10.** 14 **11.** 2 **12.** -7

13. 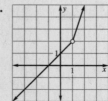 **14.**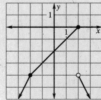

15. 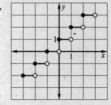 **16.**

Lesson 2.7 *continued*

17.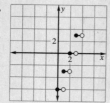

18.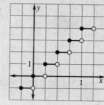

19.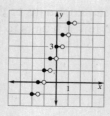

20.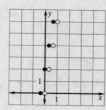

21.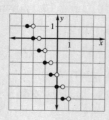

22. $f(x) = \begin{cases} 35 & \text{if } 0 \leq x \leq 6 \\ 33.80 + 0.20x & \text{if } x > 6 \end{cases}$

23.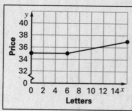

24. $36.60

25. $f(x) = \begin{cases} 15{,}000 + 0.03x, & \text{if } 0 < x < 100{,}000 \\ 18{,}000 + 0.04x, & \text{if } 100{,}000 \leq x \leq 200{,}000 \\ 20{,}000 + 0.05x, & \text{if } x > 200{,}000 \end{cases}$

26. $f(x) = \begin{cases} x, & \text{if } x \geq 0 \\ -x, & \text{if } x < 0 \end{cases}$

27. $f(x) = \begin{cases} -x - 5, & \text{if } x < -3 \\ x + 1, & \text{if } x \geq -3 \end{cases}$

Reteaching with Practice

1. -10 **2.** -2 **3.** 4 **4.** -4

5.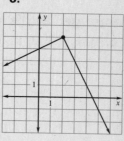

6.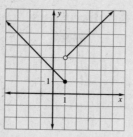

7.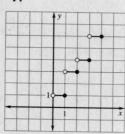

8.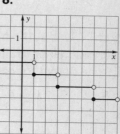

Interdisciplinary Application

1. $\begin{cases} 1.25x & \text{if } x < 10 \\ x & \text{if } 10 \leq x \leq 25 \\ 0.75x & \text{if } x > 25 \end{cases}$

2. $\begin{cases} 11.25 & \text{if } x < 10 \\ 18.00 & \text{if } 10 \leq x \leq 25 \\ 25.00 & \text{if } x > 25 \end{cases}$

3. $\begin{cases} 2x & \text{if } x < 10 \\ 20 + 1.50(x - 10) & \text{if } 10 \leq x \leq 25 \\ 42.50 + 1.10(x - 25) & \text{if } x > 25 \end{cases}$

4.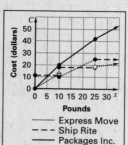

—— Express Move
– – – Ship Rite
—— Packages Inc.

5. Express Move; $17.00   **6.** For packages up to 18 pounds, use Express Move. For packages above 18 pounds, use Ship Rite.   **7.** Never

Lesson 2.7 *continued*

Challenge: Skills and Applications

1. a. 2, 1, 1, 1, 2, 1   **b.**

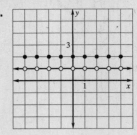

2. a. 1, 1, 0, 1, 0   **b.** Since every real number is either rational or irrational and not both, there is only one output for each input; no.

3. a. 5   **b.** $\frac{7}{6}$ **4. a.** 2, 4, 8, 16

b.

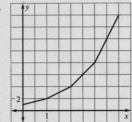

Lesson 2.8

Warm-Up Exercises

1. 6   **2.** -9 **3.** 11   **4.** 15.5   **5.** -7

Daily Homework Quiz

1. $4; -1$ **2.**

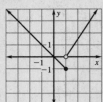

3. $f(x) = \begin{cases} 2, & \text{if } -2 \le x < -1 \\ 1, & \text{if } -1 \le x < 0 \\ 0, & \text{if } 0 \le x < 1 \\ -1, & \text{if } 1 \le x < 2 \end{cases}$

Lesson Opener

Allow 15 minutes.

Graphing Calculator Activity

1. a. shifted up 1 unit   **b.** shifted up 2 units
c. shifted down 3 units   **d.** shifted down 1 unit

2.

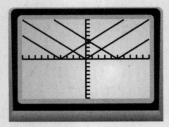

3. If h is positive, graph is shifted h units to the left. If h is negative, graph is shifted h units to the right.

Practice A

1. E   **2.** B   **3.** C   **4.** F   **5.** D   **6.** A

7. down   **8.** up   **9.** up   **10.** up   **11.** down

12. down   **13.** $(0, -3)$ **14.** $(1, 2)$

15. $(-3, -5)$ **16.** $(7, -2)$ **17.** $(-1, 9)$

18. $(-3, 0)$ **19.** same   **20.** narrower

21. wider   **22.** narrower   **23.** wider   **24.** wider

25.

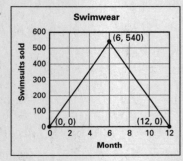

26. \$540; month number six

Practice B

1. up   **2.** down   **3.** up   **4.** $(-13, -6)$

5. $(4, -7)$ **6.** $(-2, 11)$ **7.** wider

8. narrower   **9.** narrower

10.

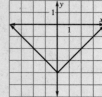

11.

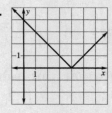

Lesson 2.8 *continued*

12.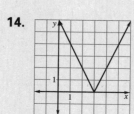

13.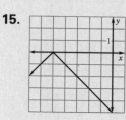

14.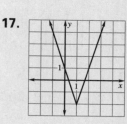

15.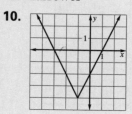

16.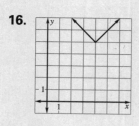

17.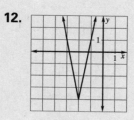

18.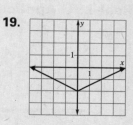

19.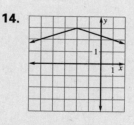

20.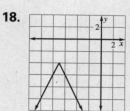

21.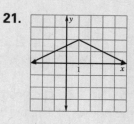

22. $y = 2|x - 3| + 1$ **23.** $y = -|x + 2| + 3$

24. $y = \frac{1}{2}|x + 2| - 1$ **25.** $(0, 22)$

26. The home is 22 feet high.

27.
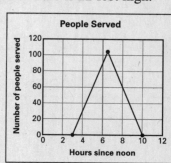

28. $(6.5, 105)$; the restaurant serves the greatest number of people, 105, at 6:30 P.M.

Practice C

1. down **2.** up **3.** down **4.** $(2, 5)$

5. $\left(\frac{3}{8}, -1\right)$ **6.** $\left(\frac{2}{3}, 6\right)$ **7.** wider **8.** wider

9. narrower

10.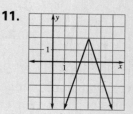

11.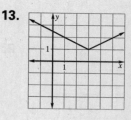

12.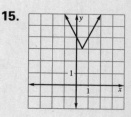

13.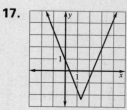

14.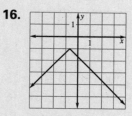

15.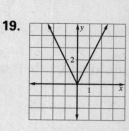

16.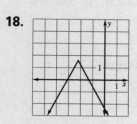

17.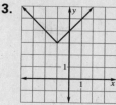

18.

19.

$y = 2|x|$

Lesson 2.8 *continued*

20.
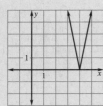
$y = 3|x|$

21.
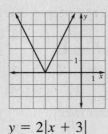
$y = 2|x + 3|$

22.
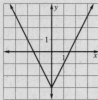
$y = 5|x - 4|$

23.
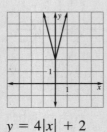
$y = 4|x| + 2$

24.
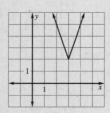
$y = 2|x| - 3$

25.
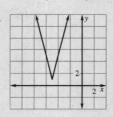
$y = 2|x + 4| + 1$

26.
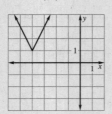
$y = 3|x - 3| + 2$

27.
$y = 4|x + 5| + 1$

28. $f(x) = \begin{cases} -50|t - 2| + 200, & \text{if } 0 \le t < 3 \\ -100|t - 5| + 350, & \text{if } 3 \le t < 6 \\ -50|t - 9| + 400, & \text{if } 6 \le t \le 10 \end{cases}$

29. $f(x) = -1.2|x| + 450$

Reteaching with Practice

1.
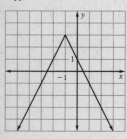

2.
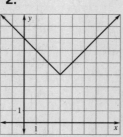

3.
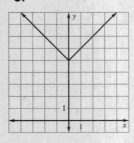

4.
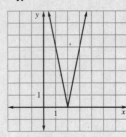

5.
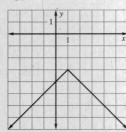

6.
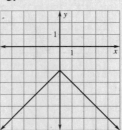

7. $y = -\frac{3}{2}|x - 3|$ **8.** $y = |x - 1| - 2$
9. $y = -4|x + 2| + 2$

Real–Life Application

1. $|x - 1|$ **2.** $|x - 0.5|, |x - 1.2|, |x - 1.3|,$
$|x - 1.8|, |x - 2|, |x - 2.5|$

3. $y = |x - 1| + |x - 0.5| + |x - 1.2| +$
$|x - 1.3| + |x - 1.8| + |x - 2| + |x - 2.5|$

4.
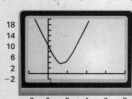

5. No   **6.** The position midway between
the houses at 1.3 and 1.8 would minimize the
distance. Therefore, the station should be located
at the point 1.55.

Challenge: Skills and Applications

1.
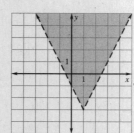

2.
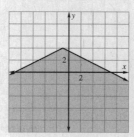

Lesson 2.8 *continued*

3.
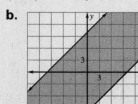

4.
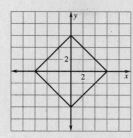

5. $h(ah + 2k)$ **6. a.** $|x - y|$ 10

b.
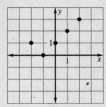

7. a. $y = \frac{5}{4}|x - 32|$;
$y = 20|x - 66|$

b. 34; it is one half your height.

Review and Assessment

Review Games and Activities

Answers will vary.

Test A

1.
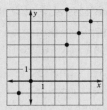

The relation is a function.

2.
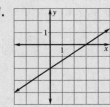

The relation is *not* a function.

3. 0   **4.** 0   **5.** 3

6.
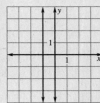

7.
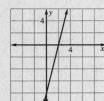

8. $y = \frac{1}{2}x$ **9.** $y = 2x + 1$ **10.** $y = x - 1$
11. $y = x - 1$ **12.** $y = -2x - 3$

13.
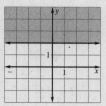

14.
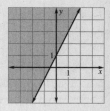

15.
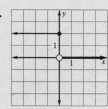

16.
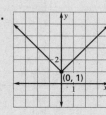

17.
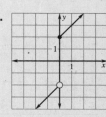

(0, 1)

18.
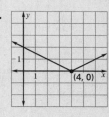

(4, 0)

19. $2s + 4a = 2800$

Test B

1.
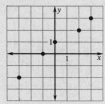

The relation is a function.

2.
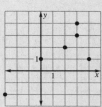

The relation is *not* a function.

3. 17   **4.** 3   **5.** 7

6.
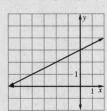

7.
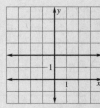

8.
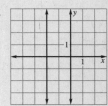

9.

10. $y = \frac{1}{2}x + 2$ **11.** $y = x - 5$ **12.** $y = x + 1$

13. $y = x + 5$ **14.** $y = -\frac{1}{2}x + 2$

Review and Assessment *continued*

15. 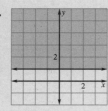 **16.**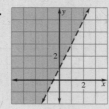

17. 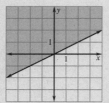 **18.**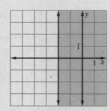

19. 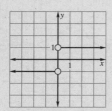 **20.**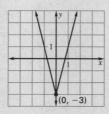

21. $n \geq 300$ **22.** (a) $\frac{3}{4}$ (b) 20 feet

Test C

1. 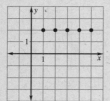 **2.**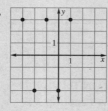

The relation is a function.   The relation is a function.

3. 5   **4.** -1 **5.** -7

6. 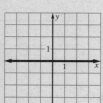 **7.**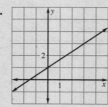

8. $y = \frac{3}{4}x - 2$ **9.** $y = -\frac{1}{2}x - 5$

10. $y = x + 5$ **11.** $y = x$ **12.** $y = -2x - 8$

13.  **14.**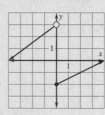

15. 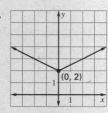 **16.**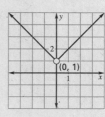

17. 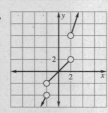 **18.**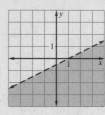

19. $8w + 10x = 3100$; 150

SAT/ACT

1. D   **2.** B   **3.** D   **4.** C   **5.** A   **6.** C   **7.** D
8. A   **9.** B   **10.** B

Alternative Assessment

1. Complete journal answers should include the following information:

a. No x-value in the ordered pairs corresponds to 2 different y-values.   **b.** Since no x-values correspond to 2 different y-values, then any vertical line should pass through the graph in *at most* one place.   **c.** When substituting x-values into the equation, each x-value yields at most one y-value.   **d.** *Sample answers*:

a. $(3, 4)\ (-3, 5)(0, -2)\ (1, 4)$ **b.** Any graph which passes the vertical line test.

c. $y = 2x^2 + 5x - 3$ **2. a.** $(66, 76), (87, 90)$
b. $y = \frac{2}{3}x + 32$ **c.** 84   **d.** 51   **3.** $x > 96$

Project: Go, Car, Go!

1. Check that the scatter plot matches the data and that the points are basically linear. The correlation is positive because the distance the car travels increases as the height of the ramp

Review and Assessment *continued*

increases. **2.** Check computations. Make sure students choose two appropriate points from the scatter plot and that the points chosen fit the linear equation found. **3.** Check computations used to make the prediction. Make sure students interpolate. The prediction should be fairly accurate. **4.** Check computations used to make the prediction. Make sure students extrapolate. If the chosen height is fairly close to the greatest height, the prediction should be fairly accurate. The farther students extrapolate, the less accurate the predication should be. **5.** The predication from Exercise 3 should be more accurate than the prediction from Exercise 4.

Cumulative Review

1. inverse property of addition
2. associative property of addition
3. distributive property
4. 7 **5.** 31 **6.** -32 **7.** -6 **8.** 6 **9.** 7
10. 71 **11.** 49 **12.** $\frac{7}{2}$ **13.** $\frac{9}{2}$ **14.** $\frac{1}{4}$

15. 2 **16.** -6 **17.** 9 **18.** $b_1 = 2A - b_2$

19. $r = \dfrac{C}{2\pi}$

20. $x < -2$ or $x > 1$ **21.** $3 < x < 5$

22. $-4 \le x \le 2$ **23.** $x < 1$ or $x > 4$

24. $\frac{16}{3}$ or $-\frac{8}{3}$ **25.** 28 or -4 **26.** 10 or -3
27. $x > 4$ or $x < -8$ **28.** $-\frac{2}{3} \le x \le 2$
29. $-2 < x < \frac{16}{5}$ **30.** 2
31. 10 **32.** 4
33. 1 **34.** -7 **35.** 16 **36.** The relation is a function. **37.** The relation is a function.
38. The relation is *not* a function. **39.** parallel
40. perpendicular

41.
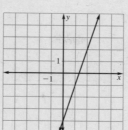

42.
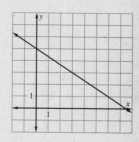

43.
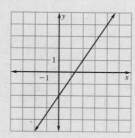

44.
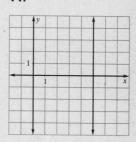

45.
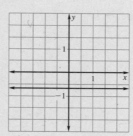

46.
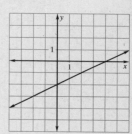

47. $y = 2x - 6$ **48.** $y = -\frac{3}{4}x + \frac{7}{4}$
49. $y = -\frac{8}{5}x + \frac{32}{5}$ **50.** $y = 1$ **51.** $x = 4$
52. $y = x + 5$ **53.** $y = \frac{7}{3}x; \frac{35}{3}$
54. $y = -2x; -10$ **55.** $y = 8x; 40$
56. $y = -\frac{1}{4}x; -\frac{5}{4}$ **57.** $y = x; 5$
58. $y = 4x; 20$

59.
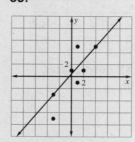

60.
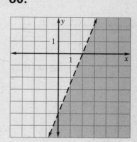

Sample answer:
$y = \frac{8}{7}x + \frac{3}{7}$

Review and Assessment *continued*

61.

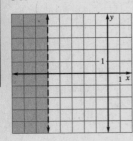

62.

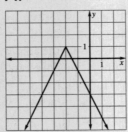

63.

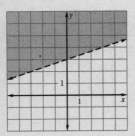

64.

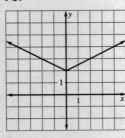

65.

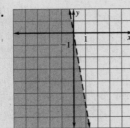

66. -1

67. 24   **68.** 5   **69.** -3 **70.** 2

71. 19

72.

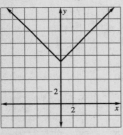

$(0, 7)$; up; same width

73.

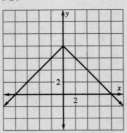

$(0, 8)$; down; same width

74.

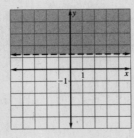

$(-2, 1)$; down; narrower

75.

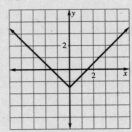

$\left(0, -\frac{3}{2}\right)$; up; same width

76.

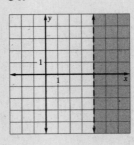

$(0, 2)$; up; wider

77.

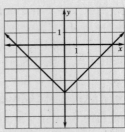

$(0, -4)$; up; same width